00-CTS-6

BEST PRACTICES FOR THE TREATMENT OF WET WEATHER WASTEWATER FLOWS

by:

Robert Brashear, Ph.D., P.E.
Cello Vitasovic, Ph.D., P.E.
Carl Johnson, P.E.
Robert Clinger, P.E.
Joel Rife, P.E.
Dan Lafitte, P.E.
Camp Dresser & McKee Inc.

Ola Nadi, Ph.D., P.E.
Nadic Engineering

2002

Water Environment Research Foundation

The Water Environment Research Foundation, a not-for profit organization, funds and manages water quality research for its subscribers through a diverse public-private partnership between municipal utilities, corporations, academia, industry, and the federal government. WERF subscribers include municipal and regional water and wastewater utilities, industrial corporations, environmental engineering firms, and others that share a commitment to cost-effective water quality solutions. WERF is dedicated to advancing science and technology addressing water quality issues as they impact water resources, the atmosphere, the lands, and quality of life.

For more information, contact:
Water Environment Research Foundation
601 Wythe Street
Alexandria, VA 22314-1994
Tel: (703) 684-2470
Fax: (703) 299-0742
www.werf.org
werf@werf.org

This report was co-published by the following organizations. For non-subscriber sales information, contact:

Water Environment Federation
601 Wythe Street
Alexandria, VA 22314-1994
Tel: (800) 666-0206
Tel: (703) 684-2452
Fax: (703) 684-2492
www.wef.org
pubs@wef.org

IWA Publishing
Alliance House, 12 Caxton Street
London SW1H 0QS, United Kingdom
Tel: +44 (0) 20 7654 5500
Fax: +44 (0) 20 7654 5555
www.iwapublishing.com
publications@iwap.co.uk

Library of Congress Catalog Card Number: 2002112604

WERF ISBN: 978-1-893664-57-9 IWAP ISBN: 978-1-84339-653-6 WEF ISBN: 978-1-57278-208-2

Camp Dresser & McKee Inc.

The research on which this report is based was funded, in part, by the United States Environmental Protection Agency through Cooperative Agreement No. CR-827345-01 with the Water Environment Research Foundation (WERF). Unless an EPA logo appears on the cover, this report is a publication of WERF, not EPA. Funds awarded under the Cooperative Agreement cited above were not used for editorial services, reproduction, printing, or distribution.

ACKNOWLEDGMENTS

Report Preparation

Principal Investigator:
Robert Brashear, Ph.D., P.E.
Camp Dresser & McKee Inc.

Project Team:
Cello Vitasovic, Ph.D., P.E.
Carl Johnson, P.E.
Robert Clinger, P.E.
Joel Rife, P.E.
Dan Lafitte, P.E.
Camp Dresser & McKee Inc.

Ola Nadi, Ph.D., P.E.
Nadic Engineering

Project Subcommittee

James R. Dartez
Royce Technologies

Robert G. Butchko, P.E.
Metropolitan St. Louis Sewer District

Carol L. Hufnagel, P.E.
Tetra Tech Inc.

Joseph A. Husband, P.E., DEE
Malcolm Pirnie, Inc.

Anthony Tafuri, P.E.
U.S. Environmental Protection Agency

David R. Williams, P.E.
East Bay Municipal Utility District

Water Environment Research Foundation Staff

Research Program Director: Jeff C. Moeller, P.E.

ABSTRACT AND BENEFITS

Abstract —
During wet weather events, separate and combined wastewater collection systems deliver substantial amounts of storm water runoff to wastewater treatment plants. Often these flows inundate collection systems and treatment works, creating bypasses of untreated or partially treated wastewater. The federal Clean Water Act (CWA), its amendments, and associated regulations have been attempting to address these concerns for nearly 30 years. These regulations, coupled with the wastewater treatment standards under the CWA, are posing immense challenges to the owners/operators of wastewater systems, who are also driven to provide the most cost-effective service to customers.

This project undertook a review of currently available technologies to improve the performance and efficiency of wet weather wastewater treatment and also identified potentially beneficial technologies and methodologies that are emerging in this area.

The project report includes the types and characteristics of current technologies and methodologies available to ensure treatment of wet weather wastewater flows, including vortex separation, enhanced clarification, operational enhancements, flushing systems, and disinfection.

Benefits —

- Identifies common wet weather wastewater treatment technologies available for use.
- Indicates applicability of each technology to types of wet weather flows.
- Indicates relative capital cost of each technology.
- Indicates relative operation and maintenance requirements and cost for each technology.
- Identifies emerging areas of practice with potential for cost-effective improvements in wet weather wastewater treatment.

Key Words —wastewater treatment, clarification, separation, operations, disinfection, flushing, stormwater, collection systems

TABLE OF CONTENTS

LIST OF TABLES

LIST OF FIGURES

LIST OF ACRONYMS

AFNOR	Association Francaise de Normalisation
APT	Advanced Primary Treatment
ATS	Aeration Tank Settling
BAF	Biological Aerated Filter
BAT	Best Available Technology
BCT	Best Conventional pollutant Technology
BMPs	Best Management Practices
BOD	Biochemical Oxygen Demand
CSO	Combined Sewer Overflow
DAF	Dissolved Air Flotation
DMHRF	Dual-Media High-Rate Filtration
DWF	Dry-Weather Flow
FC	Fecal Coliform.
GR	Gross Removal
I/I	Infiltration and Inflow
NPDES	National Pollutant Discharge Elimination System
NR	Net Removal
POTW	Publicly Owned Treatment Works
RAS	Recycle Activated Sludge
RTC	Real Time Control
SOR	Surface Overflow Rate
SS	Settleable Solids
SSO	Sanitary Sewer Overflows
TF	Treatment Factor
TKN	Total Kjeldahl Nitrogen
TMDL	Total Maximum Daily Load
TSS	Total Suspended Solids
TRC	Total Residual Chlorine
UPM	Urban Pollution Management
UV	Ultraviolet
WLA	Wasteload Allocation
WPA	Watershed Protection Approach
WWF	Wet-Weather Flow

EXECUTIVE SUMMARY

ES.1 Introduction

For at least the last 50 years, wet weather wastewater flows have created widespread difficulties. During wet weather events, wastewater collection systems often, by design or through defects, deliver substantial amounts of storm water runoff to wastewater treatment plants. Often these flows inundate collection systems and treatment works creating bypasses of untreated or partially treated wastewater.

The federal Clean Water Act (CWA), its amendments, and associated regulations have been attempting to address these concerns for nearly 30 years. Regulations and policies have been promulgated to reduce the pollution entering waterways from combined sewer overflows (CSOs), sanitary sewer overflows (SSOs), and bypasses at treatment works. These regulations and policies, coupled with the wastewater treatment standards under the CWA, are posing immense challenges to the owner/operators of wastewater systems, who are also driven to provide the most cost-effective service to customers.

This project reviews available technologies to improve the performance and efficiency of wet weather wastewater treatment and identifies potentially beneficial technologies and methodologies emerging in this area.

ES.2 About the Report

This report assessed the following categories of technologies:

- Vortex separation
- Enhanced clarification
- Operational enhancements
- Flushing systems
- Disinfection

The reader is directed to WERF report, *Assessment of Technologies for Screening, Floatable Control, and Screenings Handling* (stock no. 00CTS4) for additional information on these wet weather practices.

ES.3 Vortex Separation

Vortex concentrators are physical treatment devices that promote the settling of solids from wet weather flows. They are often referred to as "swirl" concentrators or separators since flows enter a circular basin tangentially to induce a swirl or vortex (they are also referred to as hydrodynamic separators). The swirling flow path imparts a centripetal force for solids in the flows, forcing them to the outside of the tank where they aggregate and velocities are slower. As mass increases to the outside of the tank, the solids settle and can be removed in underflow from the tank. Vortex separators were first documented in the early 1970s.

This report reviews several installations of vortex separators and documents vortex separator performance in the literature. Typical primary treatment at a wastewater treatment facility achieves net removals of 35% biochemical oxygen demand (BOD) and 50% total suspended solids (TSS). Evaluation of the net suspended solids removal from case studies and literature indicate that net removals from 5 to 15% are typical for vortex separators. They have been used primarily for CSOs, where this level of treatment is adequate. Occasionally, these separators have been combined with other treatment technologies to achieve a higher BOD and TSS removal rate.

As a wet weather management technology, vortex separation has been widely applied and is well understood. Operational issues and loss of treatment efficiency can typically be attributed to improper design or operation.

ES.4 Enhanced Clarification

Particularly within the past few years, there have been attempts to improve the solids removal efficiency of the clarification process by modifying clarifier flow regimes (e.g., lamella or plate settlers) or by chemical addition (e.g., coagulant, polymer, ballast) or by some combination of the two. Interest is increasing in the potential of dissolved air flotation and filtration to assist in the cost-effective treatment of wet weather flows.

This report reviews the following technologies:

- Lamella settlers
- Plate clarifiers
- Dissolved air flotation (DAF)
- Ballasted flocculation
- Filtration

All technologies referenced in this section have the potential to be used to treat wet weather flows. However, because of the relative newness of many of these technologies, there is a decided lack of full-scale applications on which to judge long-term viability. This is particularly true for stand-alone, end-of-pipe applications where difficulties with permitting, access, or remote operation create significant hurdles.

One of the main parameters considered when selecting a technology is the potential capital cost of the system, which includes the acquisition of land if necessary space is unavailable onsite. For this simple reason, the most attractive technologies are those that require the least amount of space. Therefore, the technologies that have higher surface loading rates would be preferred from a capital cost standpoint. These include sludge and sand enhanced clarification with lamella settling areas. Another important parameter typically considered is the required startup time for the systems to achieve steady state conditions to treat the sudden wet weather flow peaks. DAF and sand and magnetite enhanced clarification achieve steady-state technologies in approximately 15 to 20 minutes, which would allow for the system to be brought on line quickly to treat peak flows.

Application of the reviewed technologies will vary depending on parameters such as the local regulatory agencies' familiarity with the technology and the availability of land required. Systems with smaller footprints (higher loading rates) and with higher removal rates should continue to evolve as a preferred method to handle wet weather flows. Sand enhanced clarification plants have been constructed in France and are currently being constructed in the United States. Regulatory agencies will continue to become familiar with these technologies and their use will become more accepted as a method to reliably treat wet weather flows.

ES.4 Operational Enhancements

This section focuses predominantly on the methodologies that can be used with existing wastewater systems to achieve needed wet weather performance without expansion of existing works. Implementation of these methodologies may require installation of new or additional technologies, but the primary focus is on improving operational efficiency of the system.

There has been a great increase in the last 10 years in the investigation and application of real time control (RTC) to improve efficiencies. RTC is a practice that encompasses a wide array of technologies and techniques that are often selected based on a systems analysis. Computer models generated during a systems analysis (or derivatives thereof) are often used to operate RTC practices. RTC may be applied in the collection system, at the treatment plant, or both.

It is important to recognize that wet weather often introduces a significant disturbance into our system (treatment plant operation). During wet weather, significant increases in hydraulic loading can change the basic goals and the control objective function of an operation. Therefore, the process control strategies that were designed with the more advanced optimization goals are not necessarily relevant when the state of a system is experiencing significant disturbance (i.e., during a heavy storm, the operator will seldom worry about optimizing the power consumption).

During intense transient events such as storms, the primary objective is most often to avoid a loss of solids from the system. The operator will normally have control over the following process variables:

- Control of the influent flow pattern distribution among different biological reactors (step feed).
- Control of the recycle activated sludge (RAS) flow rate.
- Equalization of flow into the treatment plant.

The variables listed above are those most commonly used to deal with the impact of wet weather on the treatment process. The best strategy will depend on a number of factors, and in some cases several different aspects are considered in a comprehensive control approach.

ES.6 Flushing Systems

Sewage solids deposited in combined sewer systems during dry weather are major contributors to the pollution load from combined sewer overflows (CSO). Methods for cleaning accumulated sludge and debris in CSO and stormwater conveyance systems and storage tanks have emerged over the last 15 years by creating high speed flushing waves to resuspend deposited sediments.

This report reviews several installations of flushing systems and documents flushing system performance in the literature. The types of flushing systems assessed include manual flushing, tipping flusher, gate flusher, and spray flusher.

Previous studies found that the tipping flusher and flushing gate technology appear to be the most cost-effective means for flushing solids and debris from tanks. They also found the most efficient method for flushing large diameter flat depositing sewers is the flush gate technology.

ES.7 Disinfection

Wastewater effluents intended for discharge to receiving waters require disinfection. Ultraviolet (UV) radiation is the most common alternative used for disinfection after chlorine and chlorine derivatives.

There has been increasing application of UV at municipal wastewater plants, and WERF previously conducted a comparison between UV and chlorination that included a model to guide application (*Comparison of UV Radiation to Chlorination: Guidance for Achieving Optimal Performance*; stock no. D43008; Darby et al., 1995). UV presents several benefits over chlorine, including fewer safety concerns and elimination of the need to remove disinfectant residuals before discharge. UV also has a relatively fast startup time, generally requires less space, and could be housed at remote locations within the wastewater system. These attributes make it applicable to periodic wet weather flows. Conversely, the large capital investment, high power cost, long residence times required to achieve organism inactivation, and sensitivity to the amount of suspended solids/particles in flows work against UV for use on infrequent wet weather flows.

This report reviews several disinfection applications and documentation in available literature. Experience to date has shown that, considering a combination of performance, reliability, and cost, chlorination/dechlorination is generally the best choice for treating intermittent wet weather flows. UV and other technologies remain viable, though costly, and may be the better choice where site constraints limit the application of chlorine for disinfection.

CHAPTER 1.0

INTRODUCTION

1.1 Introduction to Wet Weather Flows

Wet weather wastewater flows have created widespread difficulties for at least the past 50 years. Wet weather flows generate from three sources: combined sewer collection systems, separate sanitary sewer collection systems, and the treatment works associated with them. During wet weather events, typical separate wastewater collection systems (designed to carry wastewater only) deliver substantial amounts of stormwater runoff to wastewater treatment plants (WWTPs) caused by infiltration and inflow (I/I). I/I enters collection systems through defects that inevitably occur in these systems as they age or through inappropriate connections of storm drainage into the sanitary collection system. Older sewer design standards often allowed sources of stormwater into sanitary sewers (e.g., roof drains and basement drains).

Although the majority of U.S. municipalities are served by separate collection systems, many urban areas are served by combined sewer systems, which are designed to convey both wastewater and stormwater. These combined systems send overflows to receiving waters during larger wet weather events. Originally seen as a cost-effective and environmentally safe means of handling urban drainage, experience has shown that these overflows create significant pollution problems from untreated wastewaters mixed with storm flows.

Often, wet weather flows into combined and sanitary sewer systems inundate collection systems and treatment works. This situation creates bypasses of untreated or partially treated wastewater at the treatment works and/or at overflows in the system ahead of the treatment works (commonly termed overflows). The difficulties in cost-effectively managing these transient flows have been recognized for some time (Field and Lager, 1975; Heaney et al., 1979).

The federal Clean Water Act (CWA), its amendments, and associated regulations have been attempting to address these concerns for nearly 30 years. The U.S. Environmental Protection Agency's (USEPA) combined sewer overflow (CSO) control policy and the emerging sanitary sewer overflow (SSO) regulations require the development of control strategies for wet weather wastewater flows for permitted facilities (treatment works and contributing collection systems). These regulations, coupled with the wastewater treatment standards under the CWA, are posing immense challenges to the owner/operators of wastewater systems, who are also driven to provide the most cost-effective service to customers (EPA, 1992).

1.2 Scope of Project

This project assesses the types and characteristics of the current technologies and methodologies available to ensure treatment of wet weather wastewater flows.

The project team contacted selected installations to obtain available design criteria and actual data on operating results. The data were assessed as to whether or not wet weather controls are performing as designed. The objective here was to determine if treatment objectives are being achieved. The project also identified new wet weather treatment technologies and methodologies and evaluated selected new technologies and methodologies.

The project objectives were realized through the following tasks:

- Task 1 – Confirm Target Areas of Investigation
- Task 2 – Collect Relevant Literature
- Task 3 – Identify Candidate Wet Weather Treatment Technologies
- Task 4 – Collect Performance Data
- Task 5 –Analyze Data and Technology
- Task 6 – Summarize Findings and Formulate Conclusions

Performance data (task 4) were more difficult than anticipated and despite repeated attempts to collect facility design criteria and operating data, neither could be collected for many facilities. Data analysis (task 5) was therefore limited. Furthermore, the scope of the projected precluded a robust assessment of the quality control and assurance aspects of the data that was made available where investigators did not have direct involvement with collection.

For more information on wet weather practices, readers are referred to *Assessment of Technologies for Screening, Floatable Control, and Screenings Handling* (stock no. 00CTS4).

CHAPTER 2.0

VORTEX SEPARATORS

2.1 Literature Review

There are two general types of vortex devices used to manage wet weather flows. The first, vortex separators/concentrators, are used for the removal of pollutants from wet weather flows. The second, vortex valves are used to regulate wet weather flows and, as such, can be used to implement a treatment scheme. However, vortex valves are not considered treatment practices.

Vortex separators are a physical treatment device designed to promote the settling of solids from wet weather flows (EPA, 1982, 1996a) and a representative vortex separator is shown in Figure 2-1. They are often referred to as "swirl" concentrators or separators since flows enter a circular basin tangentially to induce a swirl or vortex (they are also referred to as hydrodynamic

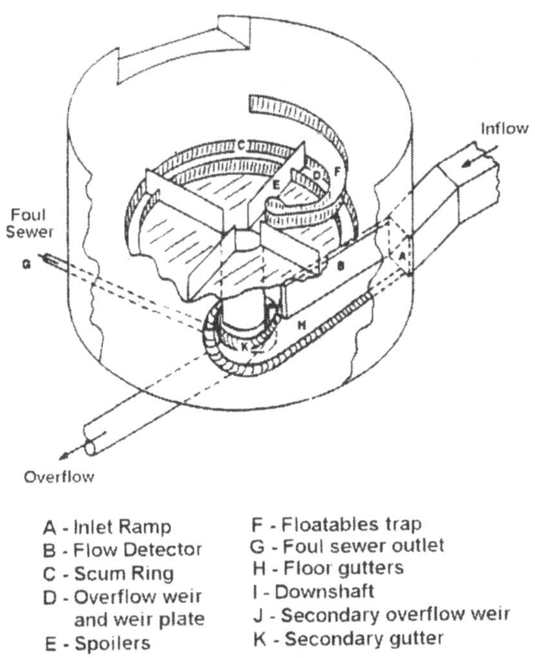

Figure 2-1: Isometric of a swirl concentrator (Source: EPA/Sullivan et al., 1982)

separators). The swirling flow path imparts a centripetal force to solids in the flows, forcing solids to the outside of the tank where they aggregate and velocities are slower. As the mass of solids increases to the outside of the tank, the solids settle and can be removed as underflow from the tank. Vortex separators can be used in a number of alternative configurations in association with storage or other technologies as shown in Figure 2-2.

Vortex separators were first documented in the early-1970s in several EPA documents (EPA, 1972, 1974, 1975, 1976, 1982, 1996a). These documents (and their associated studies) established design considerations and performance expectations. EPA's 1996 report documenting the performance of vortex solid separators for the control and treatment of wet weather flows is the most comprehensive to date (EPA, 1996a).

The development and implementation of vortex separation began in the 1970s. Despite significant application over nearly three decades, studies verifying field performance of vortex separators remains relatively limited (Schmidt et al., 1997). Furthermore, Schmidt found relatively little information on the use of vortex separators in conjunction with ancillary technologies such as storage, coagulation, process control systems, and disinfection. This literature review bears that finding out as well. While recognized as an applicable technology for wet weather wastewater flows (particularly CSOs) for a number of years, relatively little additional research has been done with the technology. The most recent and comprehensive reviews of the technology are from the USEPA (1996a) and by Field et al. (1997).

Schmidt et al. (1997) reports on the treatment efficiency of a vortex separator that was compared with a circular clarifier, a plate settler, and an inclined rotary drum screen in Toronto, Ontario, Canada. Ultraviolet (UV) disinfection was tested in the initial phase of the study to determine the effectiveness of UV irradiation of the treated effluent from the vortex separator. The study was conducted over two operational seasons and, while statistical significance of their data was not indicated, the amount of data collected was significant. The data showed suspended solids removal efficiencies achieved with the vortex separator to be equivalent to primary clarification without chemical coagulation. Although more efficient than a conventional clarifier, it was found that the relatively poor settling characteristics at the pilot plant site would require sizeable treatment units. The residual TSS concentrations were generally greater than 100 mg/L. The addition of chemical coagulation to the vortex separator improved the effluent from the vortex unit, and residual TSS concentrations substantially less than 100 mg/L were obtained. Gall et al. (1997) concurrently assessed the same facility to develop a computer model representative of the process for planning purposes.

Arnett and Boner (1998) describe the application of vortex separation on wet weather wastewater flows in Columbus, GA. Enhancement to the vortex separator was investigated including the addition of dissolved air flotation, chemical precipitation, and chemical disinfection. Performance data are provided.

Huebner and Gieger (1996) studied a system that included a vortex separator upstream of a detention facility to derive a design basis for the system. They found the system performed in the range of 10 to 35% TSS removal and that the hydraulic load was the most significant influence on dynamic settling. It was determined that to properly describe the dynamic settling as a function of the hydraulic load, the effectiveness of the settling has to be related to the mean surface load.

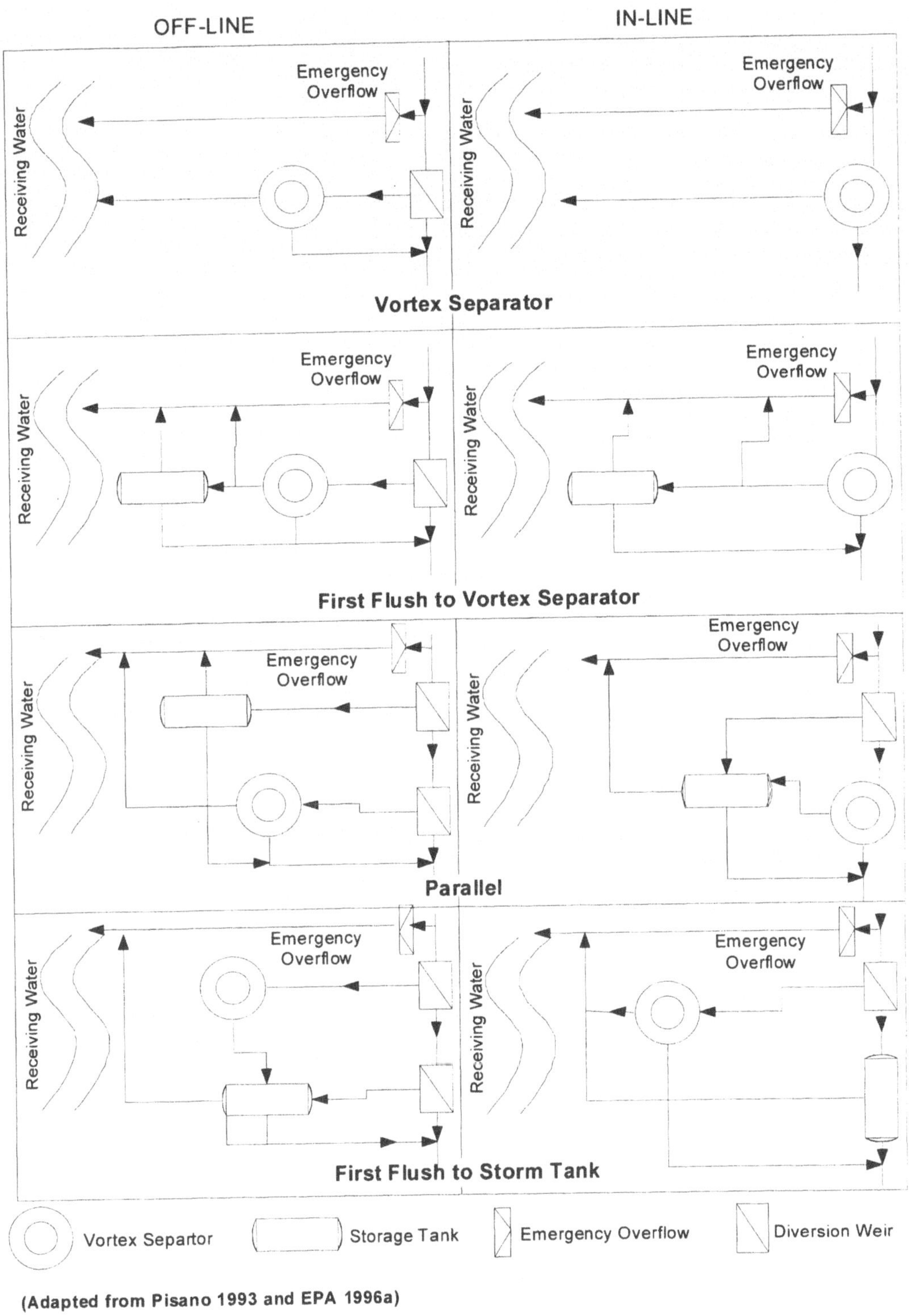

Figure 2-2: Alternative process arrangements for vortex separators (Source: Pisano, 1993 and EPA, 1996a)

EPA (1982) describes the helical bend separator for control of solids and floatables in combined sewer systems with results similar to vortex separators. The separator (illustrated in Figure 2-3) consists of a curved channel to develop helical secondary motions within the flow, capturing solids with greater settling velocities than the upward velocity of the helical flow. Unlike vortex separators, helical bend separators release solids to treatment at the end of the storm event. The separator was developed in the United Kingdom and a demonstration unit was tested in Boston, Massachusetts in the early 1980s. The helical bend separator is considerably more costly to implement than vortex separators (largely due to the considerable space required to construct).

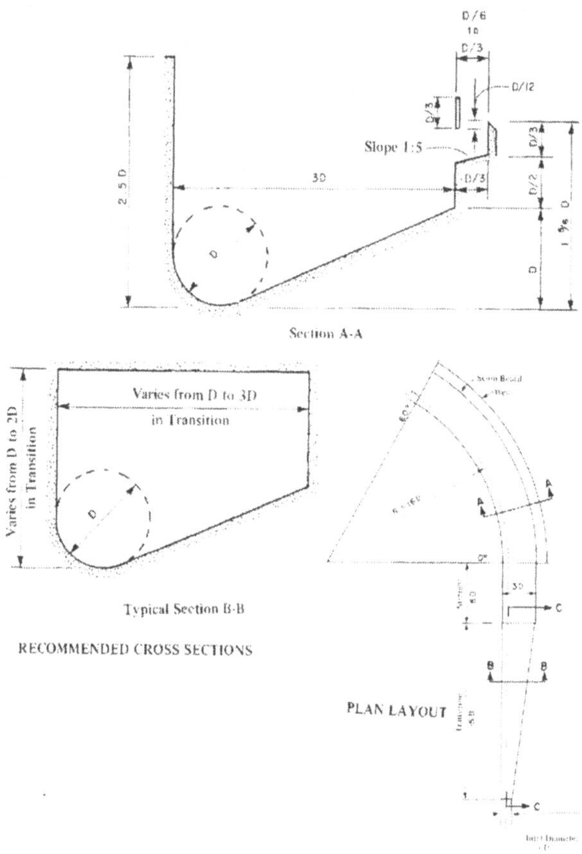

Figure 2-3: Plan layout and cross sections for helical bend separator (Source: EPA, 1982).

Applications of vortex separation evaluated by EPA in its assessment of the technology in 1996 (EPA, 1996a) point toward vortex separators as being most suited to implementation with other controls (such as detention facilities).

Several proprietary technologies using vortex separation concepts are available. These technologies are generally well documented in terms of design requirements, performance expectations, and maintenance requirements.

2.2 Performance Data

Several facilities using vortex separators are detailed below. For the evaluations in this chapter, the following equations were used to define the following terms.

$$GR_{p,t} = \left(\frac{Cin_{p,t}Vin_t - Ceff_{p,t}Veff_t}{Cin_{p,t}Vin_t}\right)$$

Gross Removal

Where:

$Cin_{p,t}$ (mg/L) — Is the flow weighted average influent concentration of parameter *p* for the duration of testing *t*

$Ceff_{p,t}$ (mg/L) — Is the flow weighted average effluent concentration of parameter *p* for the duration of testing *t*

$GR_{p,t}$ (%) — Is the gross removal efficiency for the quality parameter *p* for the duration of testing *t* otherwise defined as the mass removal efficiency

t (mins) — Is the duration of testing which can be for an entire storm event of the test equipment or for some lesser period. Where t =nΔt

Vin_t (L) — Is the volume of influent for the duration of testing *t*

$Veff_t$ (L) — Is the volume of effluent for the duration of testing *t*

$$NR_{p,t} = GR_{p,t} - \left(\frac{Vin_t - Veff_t}{Vin_t}\right)$$

Net Removal

Where:

$NR_{p,t}$ (%) — Is the net removal efficiency for the quality parameter *p* for the duration of testing *t*, otherwise defined as the differencc between mass removal efficiency and volumetric removal efficiency

$$TF_{p,t} = \left(\frac{Cu_{p,t}}{Cin_{p,t}}\right)$$

Treatment Factor

$TF_{p,t}$ — Is the treatment factor for the quality parameter *p* for the duration of testing *t*

$Cu_{p,t}$ (mg/L) — Is the flow weighted average underflow concentration of parameter *p* for the duration of testing *t*

Source: NSF, 2001.

2.3 Facilities Tested

2.3.1 Facility No. 1: DCWASA Northeast Boundary Swirl Facility, Washington, DC

This facility is located at the south end of the RFK Stadium parking lot on the west bank of the Anacostia River and adjacent to the East Side Pump Station. During storm events, this facility provides treatment and disinfection for up to 1,514,200 m^3/d (400 mgd) of combined sewer overflow before discharge to the Anacostia River.

2.3.1.1 Design criteria

The swirl was designed to provide approximately 60% reduction in settleable solids and 25 to 35% reduction in BOD.

2.3.1.2 Type of wet weather flow treated

The wet weather flow treated is CSO.

2.3.1.3 Equipment/vendor data

The facility consists of three swirls each with a 17.4 m (57 ft) diameter. The hydraulic loading rate is 2,112 $m^3/m^2/d$ (36 gpm/sq ft). The swirls were designed to operate in parallel (three units) to treat up to 1,514,200 m3/d (400 mgd) of combined sanitary sewage and stormwater. An inflatable dam is used to divert the flow to the swirl influent. When the flow becomes large enough to cause a failure of the inflatable dam, the dam will deflate, which results in the overflow of untreated CSO, until the flow has decreased sufficiently to allow the inflatable dam to resume operation. During low flow events, combined sewage is diverted to a pump station which pumps the flow to the Blue Plains treatment facility for treatment. Combined sewage in excess of this pump station's flow capacity is directed to the swirl facility. At the swirl facility, a bar screen removes floatable debris and the influent is then directed to the swirl chamber. The heavy material settles to the bottom of the chamber and is removed continuously during operation via a sewer and is sent to the Blue Plains treatment facility for treatment. The overflow from the swirl chamber is discharged via an outfall to the Anacostia River. Provision was made to both chlorinate and dechlorinate this effluent before discharge.

2.3.1.4 Installation date

The facility went into operation in January 1991.

2.3.1.5 Construction cost

Reported construction costs were $17,400,000 in 1990 dollars.

2.3.1.6 Frequency of use

Detailed information regarding the frequency of use was not provided. Frequency of use is known to be intermittent.

2.3.1.7 Wet weather flow volume treated

Maximum of 1,514,200 m3/d (400 mgd) is diverted to the swirl facility by an inflatable dam.

2.3.1.8 Influent/effluent quality

The following data represent the average concentrations for influent and effluent quality and were obtained after the Northeast Boundary Swirl Facility was first put into operation.

Table 2-1. Average concentrations for influent and effluent quality

Storm Event 1991	Total Suspended Solids (mg/L)		Volatile Suspended Solids (mg/L)		BOD (mg/L)		Settleable Solids (mg/L)	
	INF	EFF	INF	EFF	INF	EFF	INF	EFF
Mar 29	175	147	80	76	63	62	151	123
May 6	109	101	59	61	41	54	NA	NA
May 6	283	213	114	81	45	35	255	193
June 16	221	189	119	89	136	131	193	145
June 18	268	216	151	122	96	69	239	193
Mix Chamber	NA	319	NA	166	NA	128	NA	281

NA= Not available.

Source: Metropolitan Washington Council of Governments, 1991

Additional data were developed during 1992 for the Northeast Boundary Swirl Facility in an effort to document the performance of the swirl and are presented in the Table 2-2. The performance is based on suspended solids analysis. The three performance indicators are Gross Removal (GR), Net Removal (NR), and Treatment Factor (TF). If a TF is greater than 1, then the swirl is concentrating suspended solids to the underflow of the swirl. As the TF increases above 1, the swirl will improve its concentration of suspended solids.

As shown in the following data, the GR averaged 38% for all of the storm events. The data show that the NR averaged 12% with a TF of 1.46. Reviewers of the data have noted that the data results are inconsistent and care needs to be taken in the interpretation of the data. The predicted GR efficiency, estimated to be 29%, based on the average measured suspended solids settling velocities for the seven storm events, is lower than the actual measured GR of 38%, based on the average of all of the ten storm events. Previous reviewers have also noted that the data collected for the Northeast Boundary Swirl Facility also show that the swirl can remove particles lighter than the particles used for swirl development.

2.3.1.9 Method of monitoring

Samples were obtained from samplers reported to have proper intake velocities to minimize the deposition of suspended solids in the piping system. Samples were taken 45 minutes after the chemical disinfection pumps started. Difficulties with sample pumps and possible stratification of the influent in the swirl facility have raised concerns about the accuracy of the data collected during the performance evaluation.

2.3.1.10 Operating history

The swirl facility has been operational since 1991. The sampling program to demonstrate the treatment efficiency has always provided inconsistent data. The conclusions of the 1992 performance evaluation suggested that the overall gross removal was approximately 33%.

Table 2-2: Three performance indicators for the Northeast Boundary Swirl Facility 1992

Storm Event	Gross Removal of TSS (%)	Net Removal of TSS (%)	Treatment Factor (TF)	Predicted Gross Removal[a] (%)
1	28.3	13.7	1.94	16.6
3	67.5	3.1	1.05	67.0
4	39.4	5.6	1.16	--
5	27.9	13.0	1.87	22.2
7	42.0	17.8	1.73	26.0
8	35.6	24.8	3.3	10.5
9-SW2	48.9	21.8	1.83	--
10	34.9	11.0	1.46	--
11	30.0	0.0	1.00	25.7
12	21.1	8.5	1.67	32.1
Average[b]	37.6	11.9	1.46	28.6

[a] Based on settling velocity analysis

[b] Average of storm events

Source: Field and O'Connor, August 1996

2.3.1.11 Operational benefits and difficulties

Reliable data and operating efficiencies were not provided. However, DCWASA feels that the operating performance of these facilities has been inconsistent. As such, DCWASA has initiated another performance evaluation of the Northeast Boundary facility that is underway. At present, additional swirl and vortex facilities are not being considered in the CSO Long Term Control Plan being developed.

2.3.1.12 Maintenance intensiveness

Not provided.

2.3.1.13 Annual O&M cost

Not provided.

2.3.2 Facility No. 2: Tengen, Germany

The Tengen swirl facility uses a Fluidsep™ type of vortex separator. The facility consists of a parallel train of two vortex separators.

2.3.2.1 Design criteria

Each separator is 3 meters (9.8 ft) in height and diameter. The peak hydraulic design flow for each separator is 41,000 m^3/d (10.8 mgd) with a peak underflow of 1,530 m^3/d (0.4 mgd) when an overflow from the separator occurs. The underflow is discharged into a trunk sewer for final treatment at a wastewater plant. When operational with both units, the total peak hydraulic design flow is 82,000 m^3/d (21.7 mgd) and the peak underflow is 3,060 m^3/d (0.81 mgd) (which provides an underflow to inflow ratio of 0.036 at design flow conditions. The design dry weather flow was 730 m^3/d (0.2 mgd) and was later measured to be 200 m $^3/d$ (0.05 mgd) during the testing program. The sewage strength of Tengen is more concentrated than normal domestic sewage. The average total solids are 800 mg/L and the COD is 550 mg/L.

Design of a Fluidsep™ installation is based on site-specific modeling of particle settleability and includes flow and rainfall measurements.

2.3.2.2 Type of wet weather flow treated

Combined sewerage. The wet weather flow treated does not have infiltration from foundation drainage due to the installation of a parallel line which collects that flow component separately.

2.3.2.3 Equipment/vendor data

The Fluidsep™ has an aspect ration of 1 (diameter/height) that was determined after performance modeling at a German hydraulic laboratory.

2.3.2.4 Installation date

Prior to 1987.

2.3.2.5 Construction cost

Not applicable.

2.3.2.6 Frequency of use

During 1987 to 1988 hydrological year, there were 134 facility activations with 80 overflows.

Table 2-3: Overflows from 1987-1988, Tengen, Germany

	Event				
	1	2	3	4	5
Overflow duration (hr)	0.9	0.6	2.6	0.9	3.3
Underflow volume (m^3)	126	114	238	160	342
Underflow volume (as % volume)	45	39	34	82	45
Overflow volume (m^3)	152	177	740	35	427
Average surface loading during overflow ($m^3/m^2/d$)	787	1214	1181	346	653
Peak surface loading rate during overflow ($m^3/m^2/d$)	050	3523	3886	511	1807
Average underflow/inflow ratio during overflow (%)	26.9	17.4	17.9	61	41.4
Peak underflow/inflow ratio during overflow (%)	10.3	6	5.4	31.4	11.7
Source: Pisano and Brombach, 1994					

2.3.2.7 Wet weather flow volume treated

The actual volumes of wet weather treated are not documented. Listed below are flow volumes for five events. The average and peak surface loadings rates for each separator during the overflow event are also presented. Influent surface loadings during the overflow periods averaged 1,176 $m^3/m^2/d$ (20 gpm/sq ft) with a peak of 3 times the average for Events 2 and 3. The underflow rate was 194.4 $m^3/m^2/d$ (3.3 gpm/sq ft) during periods of overflow.

2.3.2.8 Influent/effluent quality

The following data represent the gross and net removal efficiencies for settleable solids for the five events. The removal efficiencies represent the percentage of the inflow constituent removed in the underflow from the process. Only simple flow splitting occurred for Event 4 as the influent hydraulic loadings were low and the relative underflow capture was high.

Table 2-4: Gross and net removal efficiencies for settleable solids for the five events.

Event	Gross removal efficiencies for settleable solids (%)	Net removal efficiencies for settleable solids (%)	Treatment factor
1	49	3.6	1.08
2	32	-7.1	0.82
3	40	15.8	1.65
4	91	9.3	1.11
5	57	12.7	1.29
Average	27	6.9	1.29
Source: Pisano and Brombach, 1994			

2.3.2.9 Method of monitoring

The samples were grab samples. Analysis of the samples included total and volatile solids, settleable solids, COD and total phosphorus. Solids settling velocities were determined from samples taken at the same time for the underflow and overflow.

2.3.2.10 Operating history

During the four years of operation during the testing program, no untreated bypasses occurred and no failures, blockages or operational problems were noted.

2.3.2.11 Operational benefits and difficulties

Not provided.

2.3.2.12 Maintenance intensiveness

Not provided.

2.3.3 Facility No. 3: Lincoln Park CSO Treatment Facility, Decatur, Ill.

The Lincoln Park CSO Treatment Facility consists of a single first flush storage tank, sized to capture 5,035 m^3 (1.33 million gallons) of the first flush volume. After the first flush occurs, there are four German designed vortex separators (Fluidsep^{TM}) capable of treating 1,574,700 m^3/d (416 mgd) to primary treatment levels. The difference between the Fluidsep^{TM} separator and the conventional vortex separator is that additional baffles and other modifications intended to minimize turbulence and increase removal efficiency in the vortex separator are not used.

2.3.3.1 Design criteria

This facility consists of a first flush tank with a diameter of 36 m (118 ft) and a side water depth of 8.5 m (28 ft). The First Flush volume is 5,035 m^3 (1.33 million gallons) and the vortex underflow volume is 2,807 m^3 (741,600 gal) for a total tank volume of 7,842 m^3 (2,071,600 gal).

Each of the four vortex units has a diameter of 13.4 m (44 ft) and is sized to treat 393,675 m^3/d (104 mgd). The total treatment flow is 1,574,700 m^3/d (416 mgd).

2.3.3.2 Type of wet weather flow treated

The Lincoln Park facility treats combined sewer flows.

2.3.3.3 Equipment/vendor data

The vortex separators were designed to remove at least 15% of the net TSS and BOD passing through them for flows less than 1,574,700 m^3/d (416 mgd). The hydraulic loading rate for the vortexes is 2,760 m^3/m^2/d (47 gpm/sq ft). The required volume to fill the vortex units prior to an overflow from the vortexes is 4,069 m^3 (1.05 million gallons). This volume also remains in the vortexes and has to be pumped out after an event. The 3,975 m^3 (1.05 million gallons) of storage is in addition to the First Flush volume of 5,035 m^3 (1.33 million gallons), which means that 9,010 m^3 (2.38 million gallons) of storage is available for an event before an overflow is initiated from the vortex.

2.3.3.4 Installation date

The Lincoln Park CSO Treatment Facility was constructed and put into operation in 1992.

2.3.3.5 Construction cost

The Total Project Cost for the Lincoln Park CSO facilities, including design, was $10,769,590 in 1992 dollars.

2.3.3.6 Frequency of use

Not provided.

2.3.3.7 Wet weather flow volume treated

The peak flow rates and the total influent volumes for specific events are shown in Table 2-5.. The peak flow rate occurred on 5/24/95 and was 1,030,000 m^3/d (272 mgd) or 65% of the design peak flow.

2.3.3.8 Influent/effluent quality

The average dry weather TSS and BOD concentration used in the design development for this facility was 291 and 188 mg/L. CBOD was used instead of BOD for the data analysis. The average dry weather CBOD concentration from July 1994 to July 1995 was 137 mg/L. The vortex separator overall event removal efficiencies are shown in the following table.

Table 2-5: Peak flow rates and the total influent volumes for seven events.

Event	Peak flow rate (mgd)	Total influent volume (million gallons)
8/13/94	66	1.92
8/30/94	200	8.40
11/20/94	78	3.21
4/16/95	155	3.44
5/8/95	192	8.45
5/16/95	265	12.54
5/24/95	272	8.69
Source: Sanitary District of Decatur, Lincoln Park Combined Sewer Overflow Treatment Facility, Operational Study, 1995		

As shown by the data, the TSS and CBOD removal efficiencies for the vortex separators were greater than 15% for all events with the peak flow rate at or below 65% of the design peak flow rate of 1,574,700 m^3/d (416 mgd). For the events of 8/13/94 and 11/20/94, the amount of CSO was very small when compared to the other events and the vortex separators were acting as primary clarifiers.

Table 2-6: Percent total gross removal efficiencies for TSS and CBOD.

Date	TSS gross removal efficiency (%)	CBOD gross removal efficiency (%)
8/13/94	90	73
8/30/94	26	22
11/20/94	74	62
4/16/95	49	46
5/8/95	44	28
5/16/95	28	37
5/24/95	41	24
Average	50	42
Source: Sanitary District of Decatur, Lincoln Park Combined Sewer Overflow Treatment Facility, Operational Study, 1995		

2.3.3.9 Method of monitoring

Flow paced sampling was started but found to have implementation issues due to the wide variation in flow rates during a CSO event. All samplers were then set to collect 24 different samples automatically every 5 minutes. Sample times were recorded either in the internal memory of the sampler or on a strip chart recorder. During a CSO event, the influent flow rate begins to be recorded as the units start to fill. The special influent sampler for the vortex units start to sample as soon as there is 25 cm (10 in) of CSO in the bottom of the structure. When the CSO fills the vortex units and begins to overflow the effluent weir, the effluent sampling starts. All data collection continues until the event is over. Samples are analyzed for TSS, VSS,

CBOD, and pH, and in some cases settleability and fecal coliform tests are also conducted. All laboratory analyses were performed in accordance with 40 CFR Part 136, Test Procedures for The Analysis of Pollutants.

2.3.3.10 Operating history

The system has operated well since startup in 1993.

2.3.3.11 Operational benefits and difficulties

No floatable material has been observed in the effluent since startup. The facility is reliable and does not require operator attendance during an event. When small CSO events occur and there is not enough flow to fill and discharge from the vortex units, attention has to be given to empty the units and clean them before the next storm event. If this is not done, then the First Flush feature of this facility becomes unusable.

2.3.3.12 Maintenance intensiveness

The facility is easy to clean. The vortex units and the first flush tank have self cleaning features incorporated into their design. The screen, distribution structures and conduits can be cleaned easily with water hoses.

2.3.3.13 Annual O&M cost

Not provided.

2.3.4 Facility No. 4: Uptown Park Water Resources Facility or Advanced Demonstration Facility, Columbus, GA

This demonstration facility was constructed in 1995. Collection of process performance data started in February 1996. The processes to be evaluated included the following:

- Hydraulic controls for floating gradient and split stream operations
- Vortex separation for heavy solids removal
- Dissolved air flotation within the vortex vessel for control of lighter particles, oil and grease
- Degriting of the vortex underflow for the protection of the downstream system
- Chemical precipitation in the vortex vessel
- High rate compressed media filtration for TSS removal and particle size control
- Chemical disinfection in the vortex vessel using sodium hypochlorite, chlorine dioxide and peracetic acid
- Decholorination using sodium bisulfite
- UV irradiation disinfection

Only the data pertaining to the operation of the vortex units are presented in this chapter.

2.3.4.1 Design criteria

Six vortex separators [9.75 m (32 ft) diameter] were constructed to receive coarsely screened CSO. The vortex separators are units provided by H.I.L. Technology, Inc. The hydraulic loading rates were up to 912 $m^3/m^2/d$ (15.5 gpm/sq ft). The flow configuration of the six units can be arranged in a parallel operation which allows for a direct treatment mode and a fill and treat mode. The facility can be set up to process the permit required 181,700 m^3/d (48 mgd) and can be operated at rates up to 545,100 m^3/d (144 mgd).

2.3.4.2 Type of wet weather flow treated

The type of wet weather flow treated is CSO.

2.3.4.3 Equipment/vendor data

H.I.L. Technology, Inc. provided the six diameter vortex separators.

2.3.4.4 Installation date

The facility was constructed in 1995.

2.3.4.5 Construction cost

Not provided.

2.3.4.6 Frequency of use

Not provided.

2.3.4.7 Wet weather flow volume treated

Not provided.

2.3.4.8 Influent/effluent quality

The overall performance of the vortex units for the first 10 monitored events during which 42 process events were documented indicated that they were achieving 40 to 60% gross TSS removal. The hydraulic loading rates during this period were as high as 912 $m^3/m^2/d$ (15.5 gpm/sq ft). The majority of the data points were at rates of 118 to 353 $m^3/m^2/d$ (2 to 6 gpm/sq ft). During larger volume events the data indicated that the percentage removals were reduced. The data demonstrate the units were achieving primary treatment removals.

2.3.4.9 Method of monitoring

The facilities are designed to monitor flow and water levels and provide for sampling around each treatment process. There are 18 automatic sampling stations provided at the influent and effluent of each treatment component. The TSS measurements are performed in accordance with EPA 160.2.

2.3.4.10 Operating history

The facility has been operational since 1996.

2.3.4.11 Operational benefits and difficulties

The Columbus Water Works (Columbus Water Works, 2001) conducted a review of some of their wet weather management strategies. Because of that review, several basic observations, conclusions, and recommendations resulting from the performance evaluation of CSO technologies in Columbus were made:

♦ One-half the number of vortex separators and twice the compressed media filter capacity would produce similar annual removals and disinfection at about two-thirds of the capital cost of the system implemented for Columbus.

♦ The optimized facility is comparable to or better than primary clarification and disinfection on an annual basis and is approximately one-half the cost that was estimated by EPA for the presumptive approach of the EPA Policy. This is due in part to, good design, the use of chemical precipitation, and operational attention to optimizing the process.

♦ The CSO facilities can be successfully operated automatically with equipment being initiated from rainfall, flow and level measuring devices.

2.3.4.12 Maintenance intensiveness

The major maintenance issue for this facility involves screenings and grit handling. A 2001 review (Columbus Water Works, 2001), recommended that future designs should include hydraulic and mechanical means to keep grit moving to the point of removal. Grit and other gross solids travel in slugs and can overwhelm undersized screening and grit handling systems.

2.3.4.13 Annual O&M cost

Not provided.

2.4 Technology Evaluation

2.4.1 Applications

An overall ranking of wet weather treatment performance factors for vortex separators is provided in Table 2-7. Each of these factors is discussed in more detail in the paragraphs that follow. The ranking provided is a general assessment based on the evaluation of the technology and, as with all wet weather treatment approaches, the performance of vortex separation is very dependent on the specific characteristics of the technology's application.

Table 2-7: Overall Ranking of Vortex Separators for Various Applications

Application	Performance Factor							
	Cost		Effort				Intensive- ness	
	Construc-tion	Annual O&M	Start-up	Shutdown & cleanup effort	Treatment effective-ness	Imple ment-ability	Operational	Maintenance
CSO	M	L	L	L	M	M	L	L
SSO	NA	NA	NA	NA	NA	NA	NA	NA
WWT at Treatment Facilities	M	L	L	L	M	M	L	L
Urban SW Runoff	M	L	L	L	M	M	L	L

H, High; M, Medium; L,Low NA, Not Applicable. SW, stormwater, WWT, wet weather treatment.

Design conditions often vary significantly. For instance, hydraulic loading rates in the applications reviewed as a part of this study varied from 912 to 2,773 $m^3/m^2/d$ (16 to 47 gpm/sq ft). To a great extent, loadings rates were established based on performance needs related to regulatory requirements.

2.4.2 Construction Cost

As with any facility, construction cost is dependent on a variety of factors. Site constraints and instrumentation requirements are two examples of factors that can have profound impact on cost. However, in considering the cost data from the literature as well as recently constructed facilities, the following cost curve seems a reasonable guide for estimating vortex separator cost. The cost (in K$) is linearly related to the surface area of each facility. These costs do not include other site improvements or interconnecting piping. These costs do include a design fee and contingency of 35%. Costs were based on 1996 costs adjusted for inflation per ENR Construction Cost Indices.

2.4.3 O&M Cost

Operation and maintenance costs are also highly dependent on facility design and frequency of use. For instance, cleaning of the unit may be done with automatic washdown facilities. Many utilities perform routine visits to unattended stations to assure the facility is operable when needed and for a check to assure the facility has not been vandalized. It is assumed that the unit will be visited every other week (26 times per year) and the inspection visit will require two hours, including travel time. Such costs are common to all types of facilities and are not unique to vortex units.

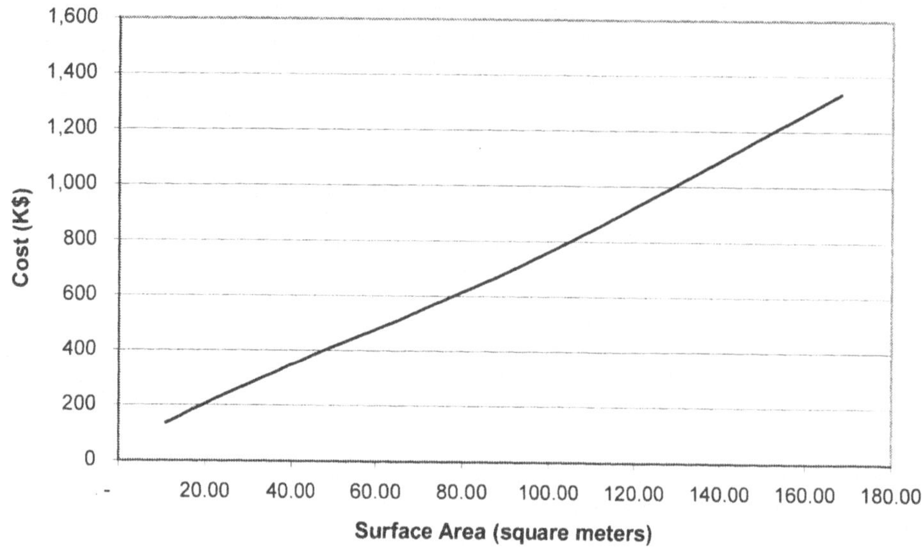

Figure 2-4: Typical cost versus surface area for vortex separators (based on EPA, 1996a)

With automatic washdown facilities only one or two special site visits to remove large objects can be anticipated. If manual hosing is required after each storm event, four to eight hours of labor per event for cleanup could be expected. Good design that eliminates opportunities for solids to settle in undesired areas (such as square corners) reduces the need and time for washdown. If a bar screen is used, frequency of cleaning will be dependent upon the normal amount of large floatables in the influent flows.

Johnson et al. (2000) reported on operation and maintenance experiences associated with CSO facilities associated with the Rouge River program in Michigan, including a facility in Redford that uses vortex separation. Johnson reported that:

- Startup and testing of any CSO basin will likely take at least two years to fine tune, given the low frequency of overflow events. Because most overflows from the combined sewer system are contained in the basins, there are only a very few events (e.g., 2 to 4) each year through which operators can develop experience with the flow-through operation of the facility.
- Design should consider operations during low flow conditions, because the vast majority of events will produce flows that are only a small fraction of the peak design flows. The use of a variety of influent pump capacities, installing flow meters to handle low flow regimes, smaller and larger screen channels, and multiple basins appear to be ways of providing the flexibility to handle low flows.

2.4.4 Startup Effort

The use of vortex separators allows the user to incorporate a treatment process that has minimal to no requirements to startup for a flow event.

2.4.5 Shutdown and Cleanup Effort

Care should be taken in the design of these facilities to allow for self-cleaning after their use. Minimal labor should have to be used if access is available to the unit and plant water or potable water is made available for washdown of the unit.

2.4.6 Treatment Effectiveness

As with any technology, treatment effectiveness is a function of design (both of the technology application, but also in system design), implementation, and operation and maintenance. Typical primary treatment at a wastewater treatment facility achieves net removals of 35% BOD reduction and 50% TSS reduction. Evaluation of the net suspended solids removal from case studies and literature indicate that the net removals from 5 to 15% are typical for vortex separators. It is apparent from the literature that variability in performance, even within the same facility, is often the case. This can be due to the variability of wet weather flows as well as to design and operation.

Field et al. (1997) felt that too little performance and operating data existed to make a quantified conclusion on expected performance. This study concurs. As previous studies have pointed out (EPA, 1996a, Field et al., 1997), the performance of vortex separation is directly related to the nature of particles and pollutants in the influent stream. Vortex separators can be expected to provide good removal of heavier particles and floatables and perform fair to poorly with lighter weight particles and material such as oil and grease, nutrients, colloidal material. Furthermore, vortex separation will generally perform less efficiently as the concentration of suspended materials in influents drop or as surface loading rates are increased.

Operational issues and loss of treatment efficiency can typically be attributed to improper design, improper flow diversion to the unit, or the improper placement of the unit such as downstream of a regulator chamber or storage chamber.

2.4.7 Implementability

The use of vortex separators has been widely accepted by regulatory agencies as part of a treatment process. Their incorporation into CSO implementation programs may be constrained by site limitations for construction of the units and ancillary components and other non-process issues. Vortex units can effectively operate in remote locations as startup operation is minimal.

Applications of vortex separation evaluated by EPA in its assessment of the technology (EPA, 1996a) and by this assessment point toward vortex separators as being most suited to implementation with other controls (such as detention facilities).

2.4.8 Operational Intensiveness

Typically, the use of a vortex separator requires minimal operator attention.

2.4.9 Maintenance Intensiveness

Maintenance procedures will have to be established for cleaning the units after each use.

2.5 Summary and Conclusions

Vortex separators have generally demonstrated that they can generally provide 5 to 15% net BOD and TSS removals. As such, their primary application is for combined sewer overflows since they are permittable under the CSO Policy. Since the effectiveness of vortex separation is generally sensitive to the particle size of solids, an assessment of solids settleability and/or pilot studies can be used to confirm performance based on local characteristics.

The performance of vortex separation seems to be influenced heavily by design and application. Proper design and proper placement of vortex separation in a wet weather treatment train can result in an effective tool for wet weather pollutant load management. As a wet weather management technology, vortex separation has been widely applied and is generally well understood.

CHAPTER 3.0

CLARIFICATION

Clarification is the process of physically removing particles from water through gravitational settling, or flotation. For the purpose of this report, clarification will also include filtration (typically referred to as a separate process, usually following clarification) since filtration is infrequently, but is occasionally used as a wet weather control method. A great deal of information in design texts is available to describe the physical phenomena associated with these processes and how they can be optimized for wastewater treatment (WEF, 1991, Metcalf & Eddy, 1991). Furthermore, design considerations for traditional sedimentation facilities used for wet weather flows are reasonably well understood (Stallard et al., 1981, Stallard et al., 1998).

A wastewater treatment facility typically includes three settling/clarification processes: grit removal, primary clarification, and secondary clarification. Each of these processes is sensitive to the hydraulic loads associated with wet weather flow. Consequently, sizing these physical processes to accommodate peak flows associated with wet weather (often 5 or more times the average daily flow) can be costly. To avoid this, sideline storage is frequently used to manage peak weather flows (Roesner and Carrier, 1996). However, storage often becomes infeasible due to site constraints or cost or both.

Particularly within the past few years, there have been attempts to improve the solids removal efficiency of clarification process by modifying clarifier flow regimes (e.g., lamella or plate settlers) or by chemical addition or by some combination thereof (EPA, 1999a). Dissolved air flotation and filtration is also being considered for treatment of wet weather flows. These processes are being used to enhance performance of existing treatment facilities, and as dedicated treatment facilities of wet weather flows located at an existing plant as a side-line process, or in the collection system at the point of overflow or for replacement of collection system storage. Prior to addressing modifications to conventional clarification, a discussion is provided regarding stress testing of existing clarifiers to establish targets for maximum loadings that should be considered prior to implementing modifications.

3.1 Conventional Clarifier Stress Testing

Ferguson et al. (2000) performed stress testing on existing primary and secondary clarifiers at two large treatment plants serving Philadelphia, Pa., as a part of a program to increase the treatment capacity of wet weather flows. Tables 3-1 and 3-2 summarize the findings.

Table 3-1.Primary Clarifier Stress Testing

Clarifier Set	Maximum SOR* $m^3/m^2/d$ (gpm/sq ft)	Maximum SOR with Modifications	Comments
Benchmark	140 (2.4)	N/A	
#1	110 (1.88)	140 (2.39)	Pretreat recycle stream from sludge processing
#2	100 (1.71)	140 (2.39)	Replace launders
#3	110 (1.88)	120 (2.05)	Increase sludge pumping
#4	85 (1.46)	140 (2.36)	Modify inlet channel and inlet openings and increase sludge pumping

*Surface Overflow Rate

Table 3-2. Secondary Clarifier Stress Testing

Clarifier Set	Maximum SOR $m^3/m^2/d$ (gpm/sq ft)	Maximum SLR* $kg/m^2/d$ (lb/sq ft/d)	Comments
Benchmark	85 (1.46)	95 (19.5)	
#1	85 (1.46)	90 (18.4)	No failure at low mixed liquor concentration
#2	65 (1.11)	165 (33.8)	Solids loading failure
#3	85 (1.46)	120 (24.6)	Combined failure
#4	85 (1.46)	80 (16.4)	No failure with gradual increase in flow
#5	85 (1.46)	75 (15.4)	No failure with sudden increase in flow

*Solids Loading Rate

Primary clarifier failure was defined as the loading that results in failure of the secondary system due to organic overload. Secondary clarifier failure was defined as the loading that results in violation of the maximum daily permit limits, 60 mg/l for both BOD and TSS. A sustainable flow analysis was also performed to determine the maximum flow rate per day that the clarifiers could be loaded at the values listed in Tables 3-1 and 3-2.

3.2 Physical Modifications

Plate and tube settlers are clarifiers, consisting of rows of inclined parallel plates or "tubes" constructed of modules of crossing inclined plates in a honeycomb pattern. As the liquid flows upward, the solids settle on the inclined, parallel plates and slides into the sludge hopper at the bottom of the unit. This has the effect of increasing the surface area of the basin allowing higher surface overflow rates and therefore providing up to 10 times the clarification capacity compared to a same footprint of a traditional clarifier. Plate and tube settlers have typically been applied to drinking water treatment processes and have not found widespread use in day-to-day wastewater treatment because of problems with plugging of the settling plates, and particularly tube settlers, by biological growth, floatables, and oil and grease.

Bridoux et al. (1998) describe the application of a plate clarification system specifically designed to treat wet weather flows at a treatment plant in France. The DELREB® system, shown in Figure 3-1, consists of two coagulation chambers, one flocculation chamber, and the plate settler with a sludge thickening zone.

Coagulants added included ferric chloride and polymer. Optimum dosages are discussed in the next section.

Coagulants added included ferric chloride and polymer. Optimum dosages are discussed in the next section. A finding of particular value in this study was that eighty percent of gross Total Suspended Solids (TSS) was removed up to a solids loading of 26 $kg/m^2/hr$ (5.3 lb/sq ft/hr), after which there was a significant drop-off of TSS gross removal. This corresponds to an extremely high hydraulic loading of 1,537 $m^3/m^2/d$ (26.4 gpm/sq ft) based on the average influent TSS of 400 mg/L. In fact, the report states that the optimum hydraulic loading was even higher at 3,600 $m^3/m^2/d$ (almost 61.1 gpm/sq ft). The ability to operate at these high hydraulic loadings were attributed to use of the appropriate coagulant dose, flocculation velocity gradient, and the sludge thickening feature in the DELREB® design.

Daligault et al. (1999) describe the performance of two-tube settler systems operated at a surface loading of 172.8 $m^3/m^2/d$ (3 gpm/sq ft) for wet weather wastewater treatment in France. Erratic performance (0–90% Settleable Solids (SS) gross removal, with an average of 54%) was attributed to plugging due to the use of tube settlers, and inadequate slope (18 degrees) of the sludge sump. It was speculated that plate settlers would have experienced less plugging problems.

The USEPA (2000) reports plans for retrofitting an existing wastewater treatment facility (the Spring Creek Water Pollution Control Facility) in New York City. A comparison is made between installing plate settlers versus chemical addition to enhance both gross and net solids removal in existing wet weather flow storage basins that discharge to Jamaica Bay. The six storage basins provide a detention time of 15 minutes for the design storm flow of 1.44 billion gallons per day. A variation of plate settlers using cross-flow (inclined plates parallel to flow) is investigated because of the success with cross-flow technology in industrial applications. Having plates parallel to the flow enhances removal of floatable as well as settleable solids, and the action of the flow across the plates tends to prevent a buildup on the plates.

Case histories discussed in the EPA report include a Shell refinery in Montreal retrofit with a stormwater holding basin. At a high overflow rate of nearly 408 $m^3/m^2/d$ (7 gpm/sq ft), TSS was reduced from 200 to 100 mg/L. Since better effluent quality was required for the Spring

Creek project, the retrofit evaluation was based on a maximum loading rate of 134 $m^3/m^2/d$ (2.3 gpm/sq ft). At this loading rate, there was only enough basin volume to accommodate around 30% of the design flow, so plate settlers were not considered feasible. Figure 3-2 is a profile view of the cross-flow plate settler used on the Shell project.

Schmidt et al. (1997) described pilot-scale studies of several combined sewer overflow technologies including horizontal flow plate clarifiers. The study demonstrated that the plate clarifiers were capable of 50% gross TSS removal and 30% gross biochemical oxygen demand (BOD) removal.

Cairo and Pujol (2000) report on the use of plate settlers in conjunction with biological aerated filters (BAF) at the Paris, France, Columbes Plant. Two operational modes are described, with the BAF being operated in series during normal flow, and in parallel during wet weather flow. The plate settlers are operated at a loading of 216 $m^3/m^2/d$ (3.7 gpm/sq ft) during normal flow, and 926 $m^3/m^2/d$ (15.8 gpm/sq ft) during wet weather flow. The efficiency of the plant during normal flow is 95% gross removal of BOD and 70% removal of nitrogen during normal flow, and 80% gross removal of BOD and 35% gross removal of ammonia during wet weather flow.

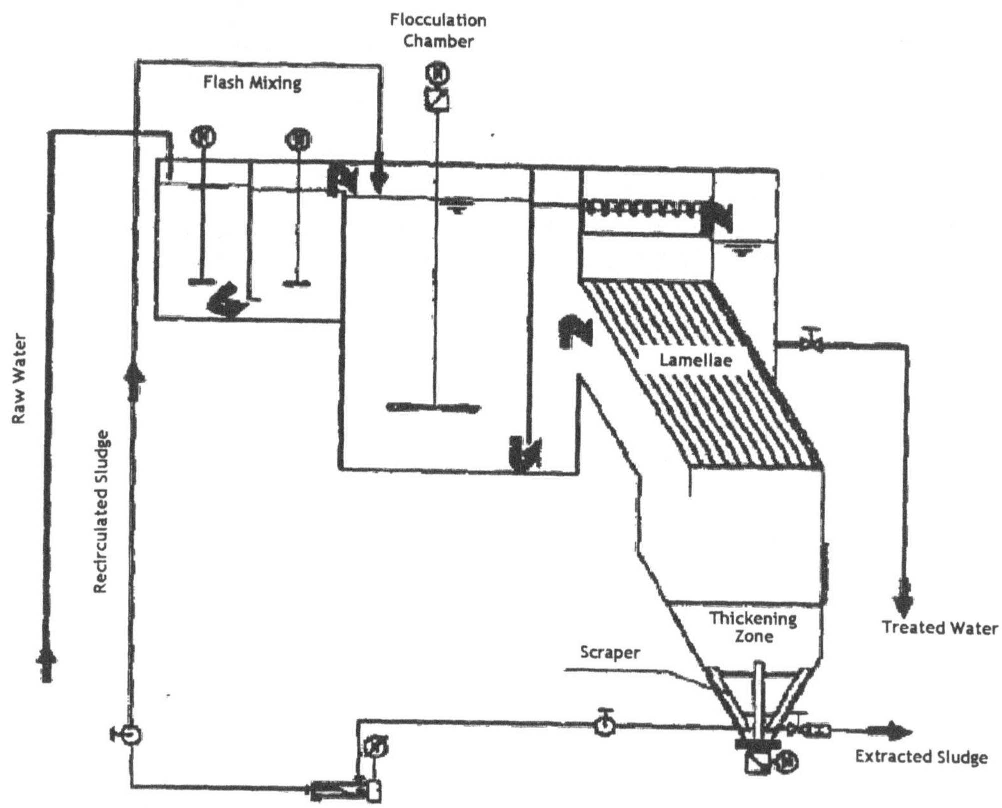

Figure 3-1: DELREB® System for Stormwater Treatment

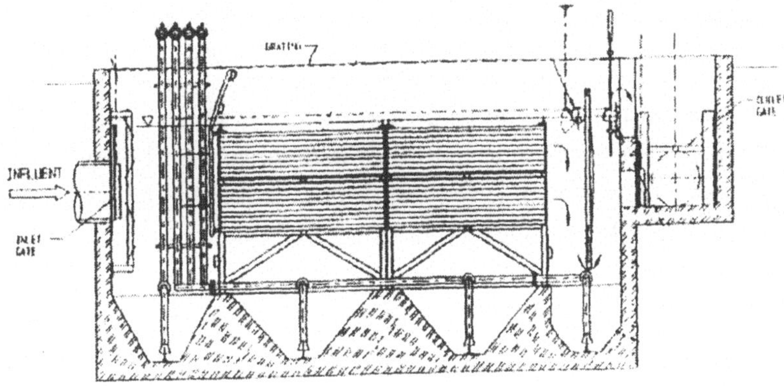

Figure 3-2: Cross-Flow Plate Settler

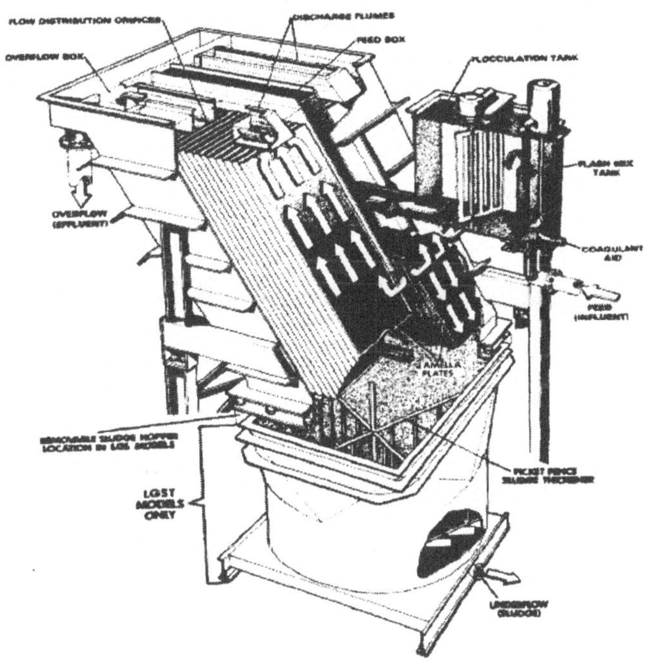

Figure 3-3: Parkson Lamella® Clarifier

West at al. (2000) investigated the use of the original plate settler, the Parkson Lamella® Plate Clarifier for treatment of stormwater. The Lamella® clarifier, shown in Figure 3-3, is the brand name frequently applied as a generic term for plate and tube settlers. The study found an unacceptably long startup time of 2 hours for the Lamella® clarifier.

3.3 Physical-Chemical Modifications

The efficiency of particle settling can be enhanced not only by physical modifications to the clarification system (e.g., plate and tube settlers), but also by modification of the particles themselves. This is typically accomplished by the addition of chemicals that promote the aggregation of particles into larger and/or heavier particles that subsequently have better settling properties. This chapter addresses (a) enhancing settling through chemical addition only, and (b) enhancing settling using a combination of chemical or weighting agent addition with optimized physical settling processes.

Bridoux et al. (1998) describe coagulation of wet weather flows at a treatment plant in France with ferric chloride and polymers. Flocculated particles were then removed by the DELREB® system shown in Figure 3-1. The optimum ferric chloride dose was found to be 0.2 mg/mg total suspended solids (TSS), or 80 mg/L based on the average influent TSS of 400 mg/L. The polymer dose was set at 3 mg/L and was not optimized.

Shao et al. (1996) describe the implementation of advanced primary treatment (APT) at the Hyperion Treatment Plant, a 350 mgd plant serving Los Angeles, Calif. When a construction project resulted in overloaded primary clarifiers 102 $m^3/m^2/d$ (1.74 gpm/sq ft), a program of application of ferric chloride and polymer was commenced. Feed rates of 10 mg/L of ferric chloride and 0.15 mg/L of anionic polymer are currently used to improve TSS gross removal from 65% to 84%, and BOD gross removal from 30% to 52%. Implementation of APT improved the operation of the downstream-activated sludge process, increasing TSS gross removal from 87 to 98% and BOD gross removal from 88% to 94%.

Heinzmann (1994) describes the application of coagulant to wet weather wastewater flows at a treatment plant in Berlin, Germany. An important point brought out in this article is that stormwater tends to be low in hardness; therefore coagulant dosages tend to be higher than in hard water. For this reason, a prepolymerized metal salt (polyaluminum chloride, or PACl) was found to be the most effective coagulant. Experiments were conducted in a 100 l/h (0.44 gpm) pilot plant. At a dose of 0.06 mmol/L Al^{3+} (5 mg/L) the 40 mg/L of influent filterable solids was reduced by 42% following sedimentation, and 87% following filtration, as compared to a reduction of only 10% in the existing settling basin without chemical addition. All heavy metals were removed "to a high degree" with the exception of zinc. It was recommended to install a system utilizing in-pipe coagulation and flocculation, settling, and filtration. Capital cost for the upgraded facilities to treat an annual average of 48,000 m^3/d (12.7 mgd) of stormwater were estimated to be $2.2 million with annual operating costs of $0.2 million.

Herr and Harper (2000) describe a low-cost method of stormwater treatment that involves flow-weighted injection of aluminum sulfate (alum) into stormwater, and settling the floc either in the receiving water, or, as the facilities are upgraded, into floc settling ponds, or in-line or in-lake floc traps. Alum treatment of stormwater has been shown to consistently reduce phosphorus by 85% to 95%, heavy metals by 80 to 90%, total nitrogen by 40% to 60%, and coliform by more than 99%. When comparing alum with ferric salts and polymer, alum was found to provide the highest pollutant gross removal efficiencies and the most stable sludge. Alum sludges have been found to age into increasing stable complexes over time, resulting in minimal dissolved aluminum concentrations.

Since the first project on Lake Ella in Tallahassee, Fla., over 30 projects have been constructed in Florida, Washington, and Indiana. For a recent project for the City of Largo, Fla., runoff from 469 hectares (1,159 acres) of urbanized land, alum is added to an existing box culvert, and the alum floc is settled out in an existing borrow pit and the alum sludge pumped into a nearby sanitary sewer. The facility is projected to remove 6,400 kg of total phosphorus, 1,600 kg (3,527 lbs) of total nitrogen, and 26,000 kg (57,320 lbs) of TSS annually for a 20-year present cost of $1.8 million, or about $1,000/kg ($545/lb) of total nitrogen. A similar size project in Hillsborough County, FL, was under construction, anticipated to remove significantly more nitrogen than the Largo project, 4,500 kg/yr, at a present worth cost of $1.4 million or $660/kg ($300/lb) total nitrogen.

Booker et al. (1996) demonstrated a proprietary high-rate clarification process for the treatment of storm induced sewer overflows capable of removing particulate solids down to at least 10 microns in size. The high-rate process involves the coagulation and flocculation of wet weather flows using iron oxide (magnetite) particles to increase floc density and particle settling rates. The coagulation process is accomplished in under two minutes, and subsequent magnetic flocculation is instantaneous. Surface loading rates on the clarifier in excess of 240 $m^3/m^2/d$ (4.1 gpm/sq ft) are achieved, and the magnetite is regenerated at high pH and recycled. Figure 3-4 is a schematic of the process.

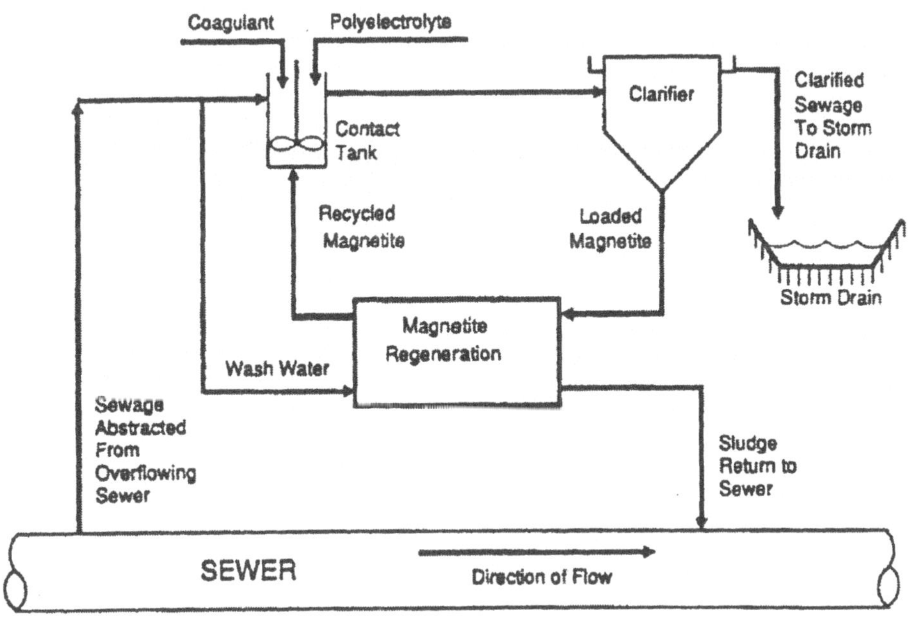

Figure 3-4: SIROFLOC® Magnetite Ballasted Coagulation Process

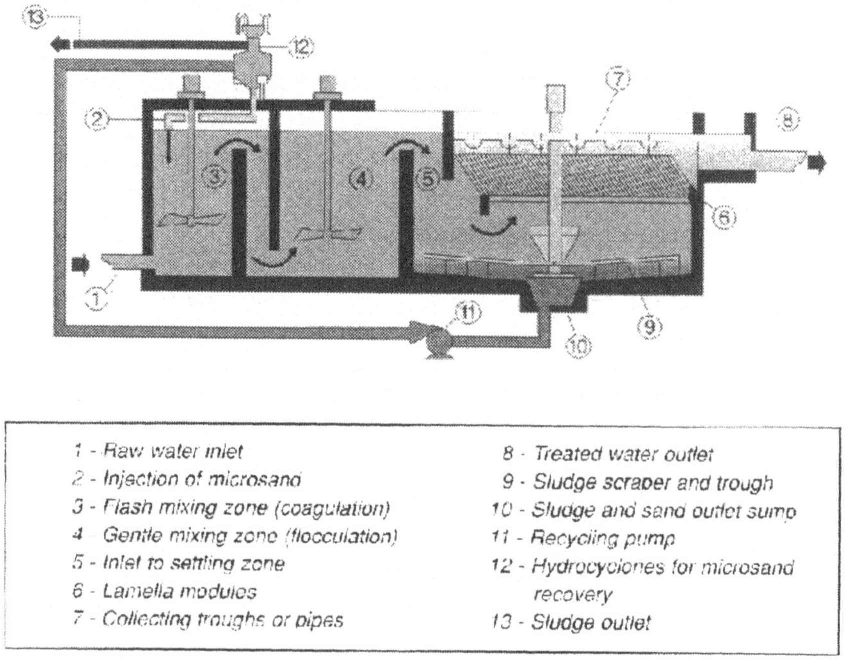

Figure 3-5: Actiflo® Ballasted Flocculation Process

In recent years, micro-sand has been used as weighted microcarrier in ballast-coagulation of colloidal particles accelerating settling velocity (EPA, 1999b). It was originally designed for drinking water treatment and is recently being tested and implemented for treating wastewater and wet weather flows. This process consists of the addition of a coagulant in the influent pipe, microcarrier and coagulant aid (i.e., flocculant, polymer, polyelectrolyte) in a mixing chamber, and then followed by maturation (flocculation) and sedimentation tanks. The initiation of the reaction of coagulation-flocculation processes is improved by the presence of the microcarrier and polymer, which increases the bonding of the floc to the microcarrier, resulting in higher settling velocities. The Actiflo® (U.S. Filter Corporation) is the only commercially available system that uses recycled microcarrier (e.g., microsand), and is frequetly referred to as ballasted flocculation processes (BFP). A variation of this process is the DensaDeg® (Ondeo Degremont Inc.) which uses recycled sludge for ballast (similar to the DELREB® process described earlier). Figure 3-5 is a section view of the Actiflo® process. A plan view of the DensaDeg® process is shown in Figure 3-6.

These processes are being evaluated at an increasing number of pilot units for treating wet weather wastewater flows. West et al. (2000) describe the pilot testing of ballasted flocculation processes (BFP) Microsep®, Actiflo®, a DensaDeg®, and lamella settlers for peak wet weather flow treatment at a wastewater plant in Fort Worth, Texas. Pilot installations of Actiflo® for wet weather wastewater treatment have shown a reduction of 85% to 95% of total suspended solids, 60 to 80% of biochemical oxygen demand and 85% to 95% of total phosphorous at overflow rates of up 10 times greater than conventional systems (West et al., 2000, Kurtz et al., 2000, Kruger, Inc., 2000). Poppe et al. (2001) report on the first high-rate clarification project in the United States in Bremerton, Ore.. A 400-gpm demonstration plant was subjected to simulated storm flow peaks and removed 90% to 95% of TSS, 80% of BOD and

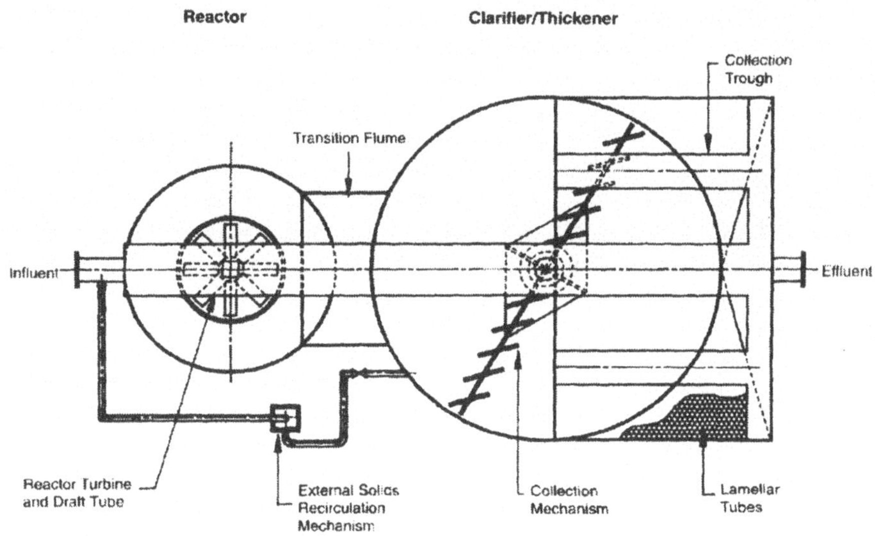

Figure 3-6: DensaDeg® Solids Contact Clarifier

85% to 90% of phosphorus, while consistently producing effluent of less than 5 NTU. A full-scale plant was started up in December 2001.

The DensaDeg® solids contact clarifier, manufactured by Ondeo Degremont, is a process utilizing settling tubes in the clarification zone in combination with sludge recirculation to a separate solids contact reactor zone. It has been compared to other high-rate clarification processes such as ballasted flocculation processes (BFP) to determine applicability for wet weather flow treatment. According to West et al. (2000), no full-scale DensaDeg® units have yet been used for treatment of wet weather influent flow. The study by West et al. reported "impressive" gross removal rates (figures not provided) but noted a startup time significantly greater than the BFP (40 to 60 minutes versus 10 to 20 minutes for the ballasted process).

The DensaDeg® clarifier was also evaluated by Kurtz et al. (2000) and compared with two ballasted flocculation processes. Primary effluent with a maximum TSS concentration of 280 mg/L was applied to the clarifier at loading rates ranging from 1,176 to 2,640 $m^3/m^2/d$ (20 to 45 gpm/sq ft). As found in the West study, performance was erratic and startup time approximately three times that of the BFP. Treatment efficiency equaled the BFP with 70% BOD and TSS gross removal, and 54% turbidity gross removal. One advantage of the DensaDeg® process is that it produces much thicker sludge than the BFP (2% to 4% versus less than 0.2% for the BFP).

EPA recently assessed a jar test methodology to evaluate high rate sedimentation processes (Ding et al., 1999). The study concluded the following:

- The use of micro-carriers (ballast) in the settling process was found to be effective in reducing coagulation-sedimentation time significantly.
- This benefit will generally extend to all storm runoff and CSO samples.
- In performing the micro-carrier-weighted jar tests developed under this study, three stages of mixing, with 10 seconds of duration in each stage, were used: (1) 150 rpm to lift the micro-carrier from the bottom of the jar into suspension, (2) 100 rpm to keep the micro-carrier in suspension, and (3) 60 rpm for the coagulation/flocculation process to treat storm

runoff samples. For CSO treatment, the third stage duration for coagulation-flocculation was found to be optimal at 1 to 2 minutes.

- The best combination of coagulant and coagulant aid for turbidity removal from surface runoff was found to be 40 mg/L aluminum sulfate (coagulant) and 1 mg/L cationic polyelectrolyte, POL-EZ-2466 (coagulant aid). For CSO, 40 mg/L coagulant (ferric chloride, as Fe+++) with 1 mg/L coagulant aid (cationic polyelectrolyte, 309C) resulted in the best turbidity removal.
- A micro-carrier particle in the range of 53 to 150 μm at a dosage of 10 g/L yielded the best results for turbidity removal from surface runoff. For CSO treatment, micro-carrier particles in the size range of 53 to 75 μm at a dosage of 3 g/L resulted in the lowest supernatant turbidity.

Each of the enhanced high-rate clarification processes generates a residual sludge, which has to be discharged from the process. Typically, the ballasted flocculation process will generate sludge at a rate of approximately 1% of throughput; therefore, the sludge produced from a 377,500 m^3/d (100 mgd) ballasted flocculation unit would be around 3,775 m^3/d (1 mgd). Sludge concentration from the ballasted process ranges from around 3,000 mg/L to 5,000 mg/L. A small percentage of micro-sand is lost from the process, which typically ranges from 1 to 2 ml/L.

3.4 Dissolved Air Flotation

Laine et al. (1998) describe an experimental system using dissolved air flotation for the treatment of wet weather wastewater flows in a separate process. This process, which combines air flotation, filtration and ultraviolet (UV) disinfection, has been tested at the pilot plant level in the eastern suburbs of Paris, France. The flotation facility has proved to be perfectly adapted to urban stormwater treatment with a maximal efficiency in the first minute of operation (90% TSS gross removal) and little sensitivity to influent pollutant load variations. An effluent TSS of 44 mg/L is reported when treating influent TSS ranging from 300 to 770 mg/L. Figure 3-7 is a schematic of the process.

EPA (1981) documented the application of dissolved air flotation to a primary treatment train and converting the secondary treatment process to contact-stabilization mode in a wastewater treatment plant serving Clatskanie, Ore., to improve wet weather performance. In wet weather flows ranging 1,893-8706 m^3/d (0.5-2.3 mgd), TSS gross removal averaged 71% and BOD gross removal averaged 73%. The capital cost of the improvements was estimated to be 14% more than a conventional treatment plant.

3.5 Filtration

The system discussed above in the Laine et al. (1998) research used a continuous backwash filter (see Figure 3-6) to further treat the DAF effluent. TSS was reduced from 44 mg/L to 20 mg/L and improved the UV transmittance to 50%.

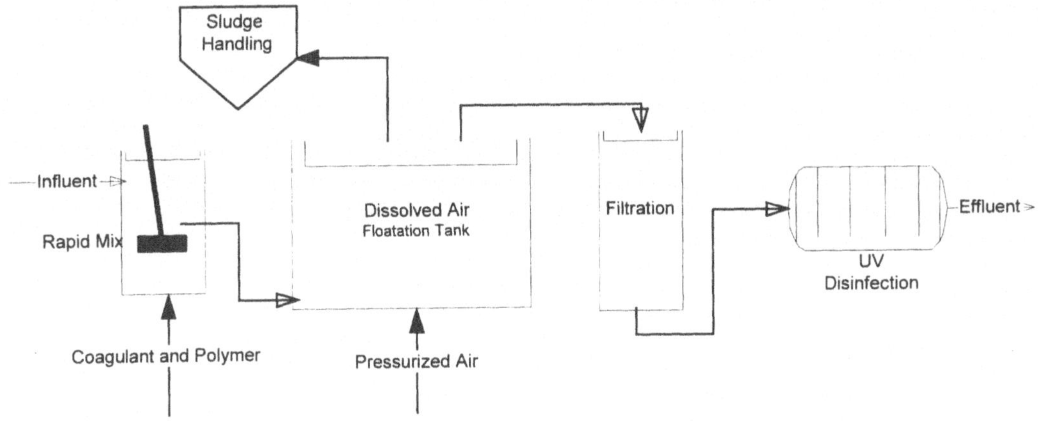

Technical Data

Dissolved Air Floatation
Height of water 0.4 m
Cell Surface 1.0 sq meters
Recirculation 40 percent
Air Flow (atm press) 0.1 cubic meters per hr
Air pressure 5.5 bars

Filtration
Height of water 4.0 m
Filtration Surface 0.7 sq meters
Particle Size 1.0 mm
Wash Water Flow 0.3 cubic meters per hr
Filtration Rate 5 m/h

Disinfection
Reactor Volume 32 liters
Number of Lamps 4
Length of Lamps 1.5 m
Germicidal Capacity 107 W
Contact Time 28 sec

Figure 3-7: DAF Process for Stormwater Treatment (adapted from Laine et al., 1998)

Booker et al. (1996) describe the application of high-rate direct filtration in conjunction with chemical coagulation (alum and polymer) and flocculation in a pilot wet weather treatment facility in Australia. The filter was loaded at a rate of 600 $m^3/m^2/d$ (10.2 gpm/sq ft). Plastic media with the least hydrophobic characteristics caused the least floc shear, resulting in a filter run time of 4 hours. Turbidity was reduced from 140 mg/l to less than 5 mg/L.

In response to a consent order to treat storm flows of twice the average daily flow, the Binghamton-Johnson City Joint Sewage Treatment Plant, New York, performed pilot studies of biological filters as reported by Aridgides, et al. (2000). Biological filters offer advantage for treating wet weather flows because the fixed biomass is less prone to washout than a suspended biomass. Influent to the filters consisted of primary effluent with an average BOD of 200 mg/L, an average TSS of 500 mg/L, and an average Total Kjeldahl Nitrogen (TKN) of 35 mg/L. The goal was to achieve permit limits of 20 mg/L of BOD and TSS and 5 mg/L total nitrogen. The biological filters were loaded at 234 $m^3/m^2/d$ (4 gpm/sq ft), twice the recommended design loading, without impairment of TSS and TKN gross removal (93% and 80%, respectively). However, BOD gross removal dropped from 88% to 57%.

3.6 Performance Data

This section summarizes performance data of the technologies discussed. Many of the technologies are in the beginning stages of development, and therefore do not have performance data or estimates of construction cost available for full-scale installations.

3.6.1 Facility No. 1: SIAAP WWTP, Paris, France – Lamella Plate Settlers with Sludge Enhanced Clarification

Bridoux et al. (1998) investigated a patented plate settler to treat wet weather flows. The intent of the study was to optimize a lamella-settling device for suspended solids gross removal and hydraulic loading. Source wastewater for the study was obtained from the SIAAP WWTP after fine screening with 5 mm openings.

3.6.1.1 Design criteria

The goals of the study were to achieve TSS gross removals in excess of 70%, hydraulic loading in excess of 2400 $m^3/m^2/d$ (41 gpm/sq ft), and a sludge concentration of greater than 3% (30 g/L). Ferric chloride and polymer were used as coagulants. Dosages, optimized by jar testing prior to the study, were 0.2 mg/L $FeCl^3$ per mg/L influent TSS, and the polymer dosage was flow-paced at a concentration of 0.8 mg/L. The ferric chloride, velocity gradient in the flocculation area and the thickened returned sludge concentrations were variables that were optimized during the study.

3.6.1.2 Type of wet weather flow treated

The screened raw sewage contained an average TSS of 400 mg/L and average COD of 700 mg/L. Flows were varied to simulate peak wet weather flows.

3.6.1.3 Equipment/vendor data

The plate settler used was designed and developed to treat wet weather flows by the Stereau Company. The DELREB® plate settler consists of two-stage flash mixing, flocculation, and a settling compartment equipped with plate settlers, a sludge thickening zone, and sludge recirculation to the flocculator (See Figure 3-1).

3.6.1.4 Installation date

The pilot plant was operated throughout much of the year 1996.

3.6.1.5 Construction cost

Cost estimates were not provided, but the DELREB® design is similar to the Actiflo® and DensaDeg® processes, so cost would be similar.

3.6.1.6 Frequency of use

The process is currently under development.

3.6.1.7 Wet weather flow volume treated

Flows were gradually increased to the process to locate the failure point. Suspended solids gross removal dropped rapidly from 80% at a solids loading of 26 $kg/m^2/hr$ (5.3 lb/sq ft/hr). Assuming a TSS influent concentration of 200 mg/L, a 20 square foot unit would be capable of treating 5792 m^3/d (1.53 mgd) of wet weather flow.

3.6.1.8 Influent/effluent quality

The parameters measured as part of the study are as follows:

Parameter	Influent (mg/L)	Effluent (mg/L)
TSS	400	120
COD	700	280

3.6.1.9 Method of monitoring

On-line TSS analyzer and COD and sludge concentration according to AFNOR (Association Francaise de Normalisation) standards.

3.6.1.10 Operating history

Product currently under development.

3.6.1.11 Operational benefits and difficulties

The very high hydraulic and solids loadings capability of this process results in a compact process. The process is also capable of producing a very thick sludge (3.5%), and can be easily automated. Difficulties would include the need for adequate preliminary treatment to prevent gross solids from causing ragging of the mixers and plugging of the plate settler.

3.6.1.12 Maintenance intensiveness

Maintenance would be dependent upon the effectiveness of the preliminary treatment system.

3.6.1.13 Annual O&M cost

Small horsepower motor for mixing and return sludge pumping would result in low electrical costs. Chemical costs would be typical for a wastewater treatment plant (assuming an influent TSS of 200 mg/L, 40 mg/L of $FeCl^3$ required and less than 1 mg/L polymer). Maintenance costs would depend on effectiveness of preliminary treatment system.

3.6.1.14 Other notes or concerns

Operating parameters that must be monitored to maintain an optimized process are coagulant dose, flocculation velocity gradient, and sludge loading. The report suggests that lamella settling with chemical addition could be used to treat wet weather flows. The surface area to treat the flows would be much less than conventional clarification units due to the high loading rates. It should be noted that these studies were carried out with influent TSS concentrations higher than typical stormwater, so reported hydraulic loading might not be possible with a more dilute influent. System should be sized based on solids loading.

3.6.2 Facility No. 2: Bremerton, Ore., CSO Treatment Facility

Based on results of successful demonstration plant, the first full scale CSO treatment facility was constructed, as reported by Poppe et al. The facility consists of a high rate clarification process utilizing the BFP technology and ultraviolet light disinfection.

3.6.2.1 Design criteria

3,216 $m^3/m^2/d$ (55 gpm/sq ft), 6 minutes total detention time. Permitted performance requirements are 55% net removal of TSS and 100% net removal of settleable solids.

3.6.2.2 Type of wet weather flow treated

Typical combined sewer overflow (50 to 100 mg/L of TSS and BOD).

3.6.2.3 Equipment/vendor data

The BFP process is the Actiflo® process by USFilter/Kruger. UV disinfection is medium pressure technology by Trojan.

3.6.2.4 Installation date

September 2001.

3.6.2.5 Construction cost

$4.1 million ($0.2/gpd treatment capacity).

3.6.2.6 Frequency of use

This is the first full-scale facility of its kind in the United States.

3.6.2.7 Wet weather flow volume treated

20 mgd.

3.6.2.8 Influent/effluent quality

Results of the demonstration plant:

Parameter	Influent (mg/L)	Effluent (mg/L)
BOD	240	45
TSS	90	9
Phosphorus	10	1

The time required to reach a consistent level of TSS net removal was 15 minutes.

3.6.2.9 Method of monitoring

Composite sampling of influent and effluent.

3.6.2.10 Operating history

Startup in December 2001. No significant rain events have occurred to date.

3.6.2.11 Operational benefits and difficulties

Benefits include:

- Quick startup time;
- Able to produce high quality effluent over a broad range of flows; and
- Small footprint.

Difficulties include:

- Sludge production of an estimated 755 m^3/d (0.2mgd) must be disposed of in the sanitary sewer; and
- Relatively high O&M cost per gallon treated due to chemical usage (note that microsand is recaptured through the use of a hydrocyclone and reused).

3.6.2.12 Maintenance intensiveness

No data to date.

3.6.2.13 Annual O&M cost

Estimated in pilot studies to be $0.09/1,000 gallons treated.

3.6.2.14 Other notes or concerns

Pilot studies carried out on simulated stormwater flows. Full-scale results are required to fully assess this technology, and the first full-scale facilities are just coming on line.

3.6.3 Facility No. 3: Village Creek WWTP, Fort Worth, Texas – Ballasted Flocculation with Sand Enhanced Clarification

West et al. (2000) investigated ballasted flocculation as a method for treating peak wet weather flows. Screened wastewater was diluted with plant effluent to simulate wet weather conditions. Full-scale design is currently under way.

3.6.3.1 Design criteria

Average Overflow Rate	2,520 $m^3/m^2/d$ (43 gpm/sq ft)
Peak Overflow Rate	3,504 $m^3/m^2/d$ (60 gpm/sq ft)
Ferric Sulfate Dose	70-125 mg/L
Polymer Dose	0.75-1 mg/L
Micro-carrier Dose	7-10,000 mg/L

3.6.3.2 Type of wet weather flow treated

Wet weather flow peaks of 3:1 caused by infiltration and inflow into the sanitary sewer system.

3.6.3.3 Equipment/vendor data

The Kruger Actiflo® process utilizes a ballast (sand) to form microfloc particles with specific gravities greater than 2.0. The Actiflo® process consists of a mixing tank, a maturation tank and a plate settler (see Figure 3-5).

3.6.3.4 Installation date

December 2003 anticipated startup date.

3.6.3.5 Construction cost

Construction costs were estimated to be \$0.2/gallon for a 377,500 m^3/d (100 mgd) facility (\$20 million).

3.6.3.6 Frequency of use

This was reported to be the first full scale application of a BFP exclusively for wet weather flow treatment

3.6.3.7 Wet weather flow volume treated

One hundred mgd is the design flow for the full-scale facility.

3.6.3.8 Influent/effluent quality

Pilot study found the following gross removal efficiencies:

Loading Rate $m^3/m^2/d$ (gpm/sq ft)	TSS Net Removal Efficiency	BOD Net Removal Efficiency
2930 (50)	89%	36%
3516 (60)	92%	64%
4102 (70)	74%	62%

The time required to reach a consistent level of TSS gross removal was approximately 20 minutes.

3.6.3.9 Method of monitoring

Testing was performed for BOD, TSS, nitrogen, phosphorus, sludge metals, sludge concentration, and startup times.

3.6.3.10 Operating history

December 2003 anticipated startup date.

3.6.3.11 Operational benefits and difficulties

The BFP has a short startup time. Chemical costs make O&M costs high.

3.6.3.12 Maintenance intensiveness

Maintenance of the BFP other than normal wear and tear is dependent on the operation of the preceding screening process.

3.6.3.13 Annual O&M cost

Estimated to be \$0.024 per m^3/d (\$90 per mgd).

3.6.4 Facility No. 4 : Hyperion WWTP, Los Angeles, Calif. – Chemically Enhanced Primary Treatment

By adding chemicals to primary clarifiers, the treatment capacity of the plant was expanded (Shao et al., 1996). While not done specifically to treat wet weather flows, this strategy could be used to address improved treatment of wet weather flow.

3.6.4.1 Design criteria

Between 10 and 40 mg/L of ferric chloride and less than 0.5 mg/L of high molecular weight anionic polymer was used to increase TSS gross removal in the primary clarifiers by 30% and BOD gross removal by 70%. By removing the BOD and TSS in the primary clarifiers, the equivalent additional treatment capacity is achieved in the secondary process.

3.6.4.2 Type of wet weather flow treated

The Hyperion project was implemented due to 25% of the primary clarifier capacity being removed from service for renovation, resulting in a simulated sustained flow peak of 1.25:1.

3.6.4.3 Equipment/vendor data

Equipment consists of chemical feeding and mixing facilities.

3.6.4.3 Installation date

The enhanced primary treatment was implemented at Hyperion in 1986 and is currently in use.

3.6.4.4 Construction cost

Capital cost for chemical feed and additional solids handling facilities were $0.015/gallon treatment capacity.

3.6.4.5 Frequency of use

Enhanced primary treatment is increasingly being considered in coastal discharge plants.

3.6.4.6 Wet weather flow volume treated

Applicability of these findings to a wet weather flow peak would be dependent upon the duration of the peak flow period. A peak flow of 11,325 m^3/d (3 mgd) for 3 hours at a 3775 m^3/d (1 mgd) average flow plant would be equivalent to a sustained peak of 1.25:1. Higher peaks would result in less of an improvement of the below stated removals.

3.6.4.7 Influent/effluent quality

net removals achieved in the advanced primary treatment:

Parameter	Influent, mg/L	Effluent mg/L
BOD	320	160
TSS	360	60

3.6.4.8 Method of monitoring

Composite sampling.

3.6.4.9 Operating history

In operation since 1986.

3.6.4.10 Operational benefits and difficulties

Operation of chemical feed and incremental increase in dewatering facilities a great deal simpler than operation of expanded activated sludge to accommodate flow peaks. Increased generation of primary sludge could increase odor problems.

3.6.4.11 Maintenance intensiveness

Maintenance of chemical feed and incremental increase in dewatering facilities much simpler than maintenance of expanded activated sludge to accommodate flow peaks.

3.6.4.12 Annual O&M cost

O&M costs for chemical feed and additional solids handling facilities were $11/mgd treated.

3.6.5 Facility No. 5: St. Barnard Parish, LA, Munster WWTP– Ballasted Flocculation Parallel Train with WWTP

The facility, in operation since November 2001, consists of ballasted flocculation units for treatment of excessive wet weather flow.

3.6.5.1 Design criteria

2,640 $m^3/m^2/d$ (45 gpm/sq ft).

3.6.5.2 Type of wet weather flow treated

Wet weather flows bypassing the biological portion of the plant.

3.6.5.3 Equipment/vendor data

Ballasted flocculation units by USFilter/Kruger.

3.6.5.4 Installation date

November 2001.

3.6.5.5 Construction cost

Wastewater plant improvements totaled $10.5 million of which 20% were related to wet weather flow treatment.

3.6.5.6 Frequency of use

This was first of this type of facility in the United States.

3.6.5.7 Wet weather flow volume treated

Flows exceeding 10 mgd are treated in the BFP. Capacity of BFP train is 10 mgd.

3.6.5.8 Influent/effluent quality

BOD influent (mg/L)	TSS influent (mg/L)	BOD effluent (mg/L)	TSS effluent (mg/L)
100-120 mg/L	100-120 mg/L	30 mg/L	10 mg/L

3.6.5.9 Method of monitoring

Composite samples.

3.6.5.10 Operational benefits and difficulties

Small footprint and short startup time. Chemical usage results in relatively high O&M costs.

3.6.5.11 Maintenance intensiveness

Dependent upon pretreatment system.

3.6.5.12 Annual O&M cost

No O&M costs to date.

3.6.5.13 Other notes or concerns

First full-scale wet weather control facility to be placed into operation in the United States.

3.6.6 Facility No. 6: Malabar Sewage Treatment Works, Sydney, Australia – Ballasted Flocculation with Magnetite Enhanced Clarification

Booker et al. (1996) investigated the use of chemical coagulation and the use of weighted particles, specifically magnetite, to adsorb coagulated particles in sewage. The magnetite is magnetically flocculated and settled out of the sewage.

3.6.6.1 Design criteria

The study intent was to illustrate that magnetite settling could allow a more compact settling device, thus reducing capital costs associated with treating wet weather events. Surface loading rates greater than 240 $m^3/m^2/d$ (4.1 gpm/sq ft) were used.

3.6.6.2 Type of wet weather flow treated

The study was conducted during normal flow conditions using screened wastewater from the Malabar Sewage Treatments Works, with flows varied to simulate wet weather flows.

3.6.6.3 Equipment/vendor data

The first unit in the Sirofloc sewage treatment process is a contact tank. Coagulant, polyelectrolyte, and magnetite are added and mixed in the contact tank, and flow continues to a clarifier. Solids from the clarifier flow to a magnetite regeneration unit recover and return magnetite to the contact tank.

3.6.6.4 Installation date

No full-scale installations to date.

3.6.6.5 Construction cost

Construction costs would be similar to other ballasted flocculation processes.

3.6.6.6 Frequency of use

No full-scale installations to date.

3.6.6.7 Wet weather flow volume treated

Not applicable.

3.6.6.8 Influent/effluent quality

Below are the results of a 5 m^3/d (1.32 mgd) demonstration plant:

Contaminant	Influent Concentration, mg/L	Gross Removal Efficiency %
TSS, mg/L	260	88
Turbidity, NTU	140	85
Oil and Grease, mg/L	60	90
COD, mg/L	630	54
Total P, mg/L	7.3	89
Total N, mg/L	33	18
Ammonia, mg/L	23	0
Total Coliforms, #/100	8×10^7	99
Copper, μg/L	286	86
Zinc, μg/L	289	85

The Sirofloc system exhibits rapid startup characteristics, producing treated sewage in less than 15 minutes.

3.6.6.9 Method of monitoring

Not reported.

3.6.6.10 Operating history

Not applicable.

3.6.6.11 Operational benefits and difficulties

The magnetite process results in very rapid coagulation (<2 minutes) and instantaneous flocculation. Problems noted include stringy flocs that do not settle well and loss of magnetite when the ends of these flocs break up. Also, polymer is not easily liberated from the floc in the regeneration process.

3.6.6.12 Maintenance intensiveness

Not applicable.

3.6.6.13 Annual O&M cost

Not applicable.

3.6.6.14 Other notes or concerns

Experimental process but has promising features for wet weather treatment.

3.6.7 Facility No. 7: CSO Outlet in Eastern Paris, France – Dissolved Air Flotation

Laine et al. (1998) investigated the use of dissolved air flotation (DAF), followed by sand filtration and ultraviolet disinfection. A pilot process was installed in eastern Paris at an existing CSO outlet, which receives flow from a combined sewer. The first process unit was a dissolved air flotation device that injects water charged with micro bubbles of air into the CSO effluent to be treated. The fine bubbles traveling through the effluent to the water surface propel the suspended solids to the surface, where a thick sludge cake was formed. The cake was then removed via a skimming device.

3.6.7.1 Design criteria

The DAF was loaded at a rate of 96 $m^3/m^2/d$ (1.6 gpm/sq ft).

3.6.7.2 Type of wet weather flow treated

The pilot process was used to treat flows from the CSO outlet during three storm events.

3.6.7.3 Equipment/vendor data

Coagulation and flocculation occurred prior to the DAF unit. Aluminum sulfate was used as a coagulant, and a Floerger brand anionic polymer was used. Effluent from the coagulation tank flowed directly to the DAF unit. Effluent from the filtration unit was used as source water for the addition of the micro bubbles into the DAF unit. Effluent from the DAF unit continued through the sand filtration column and the UV disinfection system.

3.6.7.4 Installation date

No full-scale installations to date.

3.6.7.5 Construction cost

Cost estimates not provided.

3.6.7.6 Frequency of use

No full-scale installations to date.

3.6.7.7 Wet weather flow volume treated

Variable runoff from an industrial area of 284 ha (700 acres) of which 28% was impermeable. Average flow rate was 96 m^3/d.

3.6.7.8 Influent/effluent quality

Parameter	Influent, mg/L	Effluent, mg/L
TSS	460	44
COD	600	79

3.6.7.9 Method of monitoring

Automatic composite sampling.

3.6.7.10 Operating history

Not reported.

3.6.7.11 Operational benefits and difficulties

The DAF process is a known process. It is somewhat complex mechanically, and floating sludge can be difficult to handle.

3.6.7.12 Maintenance intensiveness

Maintenance of process not provided.

3.6.7.13 Other notes or concerns

Effluent from the DAF unit achieved the net removal rates within the first minute of operation. The average output during all storm events from the DAF unit was 44 mg/l of TSS.

3.7 Technology Evaluation

An overall ranking of clarification technologies separators are provided in Tables 3-1 to 3-3.

Table 3-3: Overall Ranking of Plate Settlers for Various Applications

Application	Potential Applicability		
	BOD/TSS Gross Removal Performance	Life-Cycle Cost (Capital + O&M)	Notes
CSO	Medium	Depends on application.	The loading rates for lamella settlers are lower than other technologies. Therefore, greater capital costs would be required for this technology. Plate settlers can experience plugging problems, thus require routine cleaning
SSO	Medium	See Note for CSO	See Note for CSO
WWT Treatment Facilities	Medium to High	See Note for CSO	Plate settlers can be used to retrofit existing clarifiers, resulting in a cost-effective method of improving treatment of wet weather flow.
Urban SW Runoff	Medium	See Note for CSO	See Note for CSO

WWT, Wet weather treatment; SW, stormwater.

Table 3-4: Overall Ranking of Micro-carrier Enhanced Clarification (Ballasted Flocculation) for Various Applications

Application	Potential Applicability		
	BOD/TSS Gross Removal Performance	Life-Cycle Cost (Capital + O&M)	Notes
CSOs	High	Capital costs ~ 12% the cost of biological processes. Operating costs ~ 3x the cost of biological processes.	Both the sand ballasted clarification and the magnetite ballasted clarification had gross removal efficiencies in the upper 80%age range. However, the magnetite system has a much lower overflow rate, making it much less attractive from a capital cost standpoint.
SSOs	High	See CSO note	See CSO note
Wet Weather Treatment at Treatment Facilities	High	See CSO note	See CSO note
Urban SW Runoff	High	See CSO note	See CSO note

WWT, Wet weather treatment; SW, stormwater.

Table 3-5: Overall Ranking of Dissolved Air Flotation for Various Applications

Application	Potential Applicability		
	BOD/TSS Gross Removal Performance	Life-Cycle Cost (Capital + O&M)	Notes
CSOs	Medium	Capital cost of DAF processes are less than biological processes but greater than ballasted flocculation processes. Operating costs similar to ballasted flocculation processes.	The DAF unit alone had high gross removal rates for the CSO events treated. Loading rate is significantly lower than those associated with ballasted flocculation processes.
SSOs	Medium	See CSO note	See CSO note
WWT Treatment Facilities	Medium to High	See CSO note	A DAF process could be retrofitted into existing process tanks, lowering cost.
Urban SW Runoff	Medium	See CSO note	See CSO note

WWT, Wet weather treatment; SW, stormwater.

Further ranking of wet weather treatment performance factors for enhanced clarification technologies are listed in Table 3-6 through 3-11. For this ranking, H = high, M = medium/mid-range, L = Low, and NA = Not Applicable.

Table 3-6: Ranking of Performance Factors for Plate and Tube Settlers

Application	Performance Factor							
	Construction Cost	Annual O&M Cost	Startup Effort	Shutdown & cleanup effort	Treatment effectiveness	Implement-ability	Operational Intensiveness	Maintenance Intensiveness
CSOs	H	L	L	L	M	H	M	M
SSOs	H	L	L	L	M	L	M	M
WWT at Treatment Facilities	H	L	L	L	M	H	M	M
Urban SW Runoff	H	L	L	L	M	M	M	M

WWT, Wet weather treatment; SW, stormwater.

Table 3-7: Ranking of Detailed Performance Factors for Plate and Tube Settlers with Sludge Enhanced Clarification

Application	Performance Factor							
	Construction Cost	Annual O&M Cost	Startup Effort	Shutdown & cleanup effort	Treatment effectiveness	Implement-ability	Operational Intensiveness	Maintenance Intensiveness
CSOs	L	M	H	H	M	M	H	M
SSOs	L	M	H	H	M	L	H	M
WWT Facilities	L	M	H	H	M	M	H	M
Urban SW Runoff	L	M	H	H	M	L	H	M

WWT, Wet weather treatment; SW, stormwater.

Table 3-8: Ranking of Detailed Performance Factors for Micro-carrier Enhanced Clarification

Application	Performance Factor							
	Construction Cost	Annual O&M Cost	Startup Effort	Shutdown & cleanup effort	Treatment effectiveness	Implement-ability	Operational Intensiveness	Maintenance Intensiveness
CSOs	L	M	M	M	H	M	M	M
SSOs	L	M	M	M	H	M	M	M
WWT Facilities	L	M	M	M	H	M	M	M
Urban SW Runoff	L	M	M	M	H	M	M	M

WWT, Wet weather treatment; SW, stormwater.

Table 3-9: Ranking of Detailed Performance Factors for Dissolved Air Flotation

Application	Performance Factor							
	Construction cost	Annual O&M cost	Startup Effort	Shutdown & cleanup effort	Treatment effectiveness	Implement -ability	Operational intensiveness	Maintenance intensiveness
CSO	H	H	L	M	H	M	M	M
SSO	H	H	L	M	H	M	M	M
WWT Facilities	H	H	L	M	H	M	M	M
Urban SW Runoff	H	H	L	M	H	M	M	M

WWT, Wet weather treatment; SW, stormwater.

3.8 Summary and Conclusions

All technologies referenced in this chapter have the potential to be used to treat wet weather flows. The advantage of these technologies are their abilities to handle higher loading rates, which in turn can reduce overall cost for treating intermittent high flows.

The enhanced high-rate clarification processes are very attractive for implementation as a wet weather treatment train because their capital costs are a fraction of typical biological

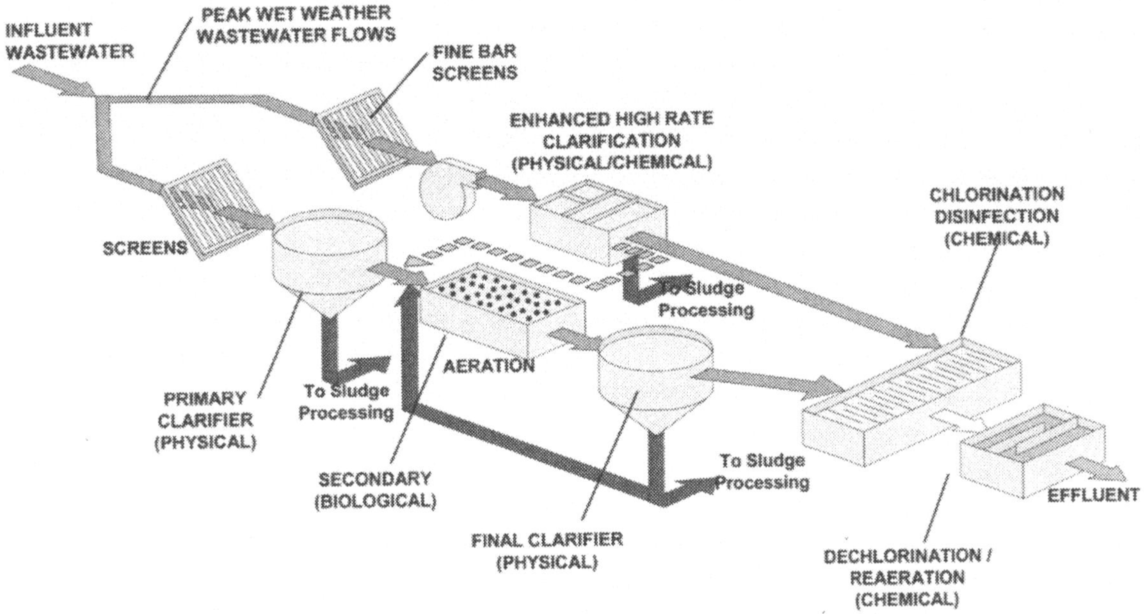

Figure 3-8. Wastewater treatment flow schematic with enhanced high-rate clarification process

treatment. While the operational expense is significant due to chemical and power costs, these expenses are generally incurred only during the infrequent peak flow event and permit compliance can be achieved. Another important parameter typically considered is the required startup time for the systems to achieve steady state conditions to treat the sudden wet weather flow peaks. Both DAF-, sand- and magnetite-enhanced clarification achieve steady state in approximately 15-20 minutes, which would allow for the system to be brought online quickly to treat peak flows.

Ideally, enhanced high-rate clarification would be integrated as a parallel train into existing and new wastewater treatment facilities to provide peak flow shaving and to allow the biological process to treat a constant flow (see Figure 3-8). In this arrangement, effluent from the enhanced high-rate clarification unit, prior to the unit reaching full efficiency, would be returned to the head of the aeration basins. Once the enhanced high-rate clarification unit reaches peak efficiency, approximately 20 minutes, the effluent from the unit would be combined with activated sludge effluent, disinfected and discharged to the receiving stream. During non-peak wet weather flows, the enhanced high-rate process could be utilized as a chemically enhanced primary clarifier to help maintain flow during construction periods at the head of the plant. Also, it is possible to use the enhanced high-rate clarifiers within the collection system at permitted known combined/sanitary sewer overflows to reduce the amount of untreated discharges as well as provide a higher level of treatment (often equivalent to secondary treatment) than with standard stormwater clarifiers.

CHAPTER 4.0

FLUSHING SYSTEMS

Sewage solids deposited in combined sewer systems during dry weather are major contributors to the pollution load from combined sewer overflows (CSO). Innovative methods for cleaning accumulated sludge and debris in CSO and stormwater conveyance systems and storage tanks have emerged over the last 15 years by creating high speed flushing waves to resuspend deposited sediments (EPA, 1998a). Figure 4-1 illustrates two types of flushing mechanisms used: flushing gate and tipping flushers.

Cleansing efficiency of periodic flush waves depends on flush volume, flush discharge rate, sewer slope, sewer length, sewer flow rate, sewer diameter, and population density (EPA, 1998a). Maximum flushing volumes at upstream points are limited by available space, hydraulic limitations and costs. The regulator/interceptor capacities before overflow limits maximum flushing rates at the downstream point. The relationship between cleaning efficiency and pipe length is important. The aim of flushing is to wash the resuspended sediment to strategic locations (i.e., to a point where the waste stream is flowing with sufficient velocity, to another point where flushing will be initiated, to a storage sump which will allow later removal of the stored contents, or to the wastewater treatment plant (WWTP). This reduces the amount of solids resuspended during storm events, lessens the need for CSO treatment and sludge removal at downstream storage facilities, and allows the conveyance of more flow to the WWTP or to the drainage outlet.

Case studies of 18 sewer and tank sediment-flushing facilities were conduct by EPA (Pisano, 1998). From the case studies, it was concluded that the tipping flusher and flushing gate technologies appear to the most cost-effective means for flushing solids and debris from tanks. Flushing gate technology appeared to be the best suited for large, flat sewers. Flushing also was demonstrated to be cost-effective for reducing hydrogen sulfide generation.

Research has shown that polymeric injection can greatly increase flow capacity (by reducing wall friction) and thus be used to correct pollution causing conditions such as localized flooding and excessive overflows. Direct cost savings are realized by eliminating relief-sewer construction (Chandler and Lewis, 1977).

As a follow-up to an earlier study, providing simple equations for predicting dry-weather deposition in sewers, a report was published showing that sewer flushing flush waves can effectively convey sewer deposits including organic matter (Pisano and Queriroz, 1977, Pisano et al., 1979).

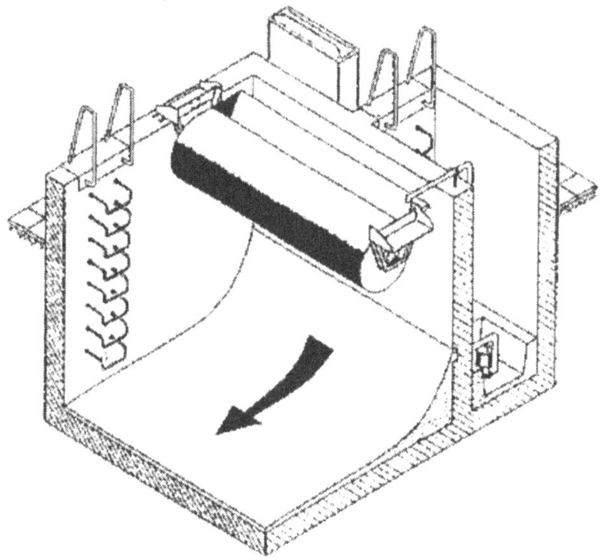

Typical Tipping Flusher Configuration

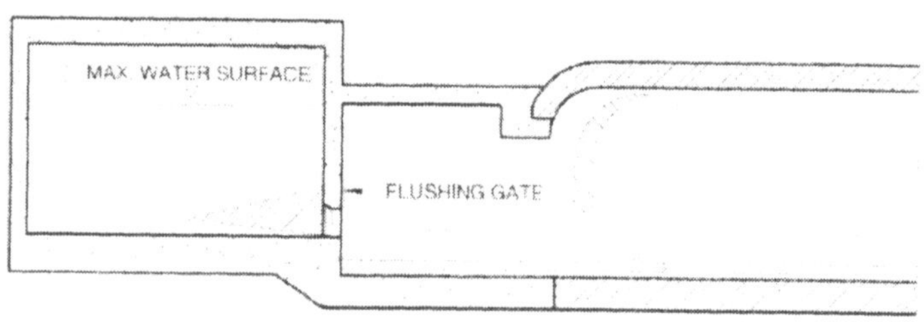

Flushing Gate Configuration (Whitten, Germany)

Figure 4-1: Tipping flusher and flushing gate configurations (Source: Pisano, 1998)

In another study, it was concluded that sewer flushing could reduce CSO control costs 7% when compared to a CSO storage/treatment and disinfections facility designed for a one-year storm (Kaufman and Lai, 1980).

Lorenzen et al. (1996) conducted flush cleaning experiments in a combined sewer system in the City of Hanover, Germany, using two existing gate valves. The objective of the experiment was to find a cleaning method and the frequency for the need of flush cleaning. The flush cleaning is a powerful method to remove the accumulated sediments in the combined sewer system. The sediment heights and water depth were measured on five-meter intervals along the experimental sewer section. Different flushing programs were tried by changing the flush water height, velocity and duration. The sediment height was measured for each program to find the removal efficiency of that particular flushing practice. The results showed that substantial amounts of sediments could be moved using gate valves located upstream of the section. In addition, for removing the sediments from the sewer pipes, it is favorable to do frequent flushes rather than a single intense flush for a long time.

Sakakibara (1996) carried out a flushing experiment in a trunk sewer. The objective of the experiment was to identify the physical and chemical characteristics of the sediments and estimate the effect of flushing in cleaning the sewer pipes. The flushing flow rate ranged between 0.40 and 0.44 m3/s in a conveyance with a maximum diameter of 6 m (19.69 ft).

Results showed that tractive force at the flat portion of trunk sewer was calculated as 0.3 N/m2; thus, the critical particle size was predicted as 0.36 mm. However, the decrease of sediments height was only 1 mm, and no difference of particle size distribution between before and after flushing. Sediments consisted of inorganic and organic substances. For organic substances, the specific gravity was 2.50, water content was 73%, and the volatile solids portion was 39%. For inorganic substances, specific gravity was 2.60, water content was 20%, and volatile solids was 1%. Sulfide concentration suggested no serious problem of odor and corrosion. The flushed load ranged from 57 to 401 kg for BOD and 108-2,960 kg for suspended solids. The concentration of flushing water ranged from 100 to 640 mg/s as suspended solids. Estimated prediction of 1% to 30% of total deposited load in the trunk sewer was removed by one flushing event. The ratio of reduced load by flushing during the course of the year to the total discharged load over a year was estimated as 4%.

4.1 Performance Data

The data presented in the subparagraphs below pertain to operating information for flushing pipelines. In defining the terms of performance (excellent, good and fair), the following benchmarks were established. Excellent meant that all sediments in channel or bay were cleaned with the flush. Good indicated substantial removal of sediments in channel, or 90% of flush lane or pipe is cleaned. Fair meant partial removal of sediments, or 50 to 70% of flush lane or channel cleaned with flush. Data, which pertained to the operation of flush gates and tipping flushers for tanks, are not included.

4.1.1 Facility No. 1: Marht Wiesentheid, Germany

This is an in-line CSO storage conduit which is throttled at its outlet by a flow regulator to maintain outflow equal to two times the average dry weather flow including infiltration.

4.1.1.1 Design criteria

The pipeline is 1800 mm (72 in), circular and 46.8 m (154 ft) in length with an average slope of 1%. An in-line flushing vault holding 14 m^3 (3,700 gal) is used to clean the pipe, and the flushing water source is local wastewater/combined sewer flows. The flush vault size is 2.5 m long by 4.84 m wide and 1.15 m high (8.2 ft long by 15.9 ft wide by 3.8 ft high). The flush gate is 2.8 m (9.2 ft) long and 1.31 m (4.3 ft) high. It is stainless steel and is controlled hydraulically. The slope of the flush vault is 20%.

4.1.1.2 Type of wet weather flow treated

Combined sewer.

4.1.1.3 Equipment/vendor data

Not applicable.

4.1.1.4 Installation date

1992.

4.1.1.15 Construction cost

Not provided.

4.1.1.6 Frequency of use

Not provided.

4.1.1.7 Wet weather flow volume treated

Not provided.

4.1.1.8 Influent/effluent quality

Not applicable.

4.1.1.9 Method of monitoring

Not provided.

4.1.1.10 Operating history

Operators note excellent performance in cleaning deposits during flushing. This facility was modeled, and the resulting terminal velocity of the sewer was calculated to be 3.1 m/s (10.2 ft/s). This value correlates with the reported performance by the operators.

4.1.1.11 Operational benefits and difficulties

None noted.

4.1.1.12 Maintenance intensiveness

There is a need for one person to inspect the facility after each activation.

4.1.1.13 Annual O&M cost

Not provided.

4.1.2 Facility No. 2: Gemeinde Schauenburg, Germany

This is an in-line CSO storage conduit which is throttled at its outlet by a flow regulator. The flushing system is used when the downstream flow regulator permits the storage to drain.

4.1.2.1 Design criteria

The pipeline is 2000 mm (78 in), circular and 64 m (210 ft) in length with an average slope of 1%. An in-line flushing vault holding 5.5 m^3 (1,450 gal) is used to clean the pipe. The flush vault size is 2 m long by 3 m wide and 0.91 m high (6.6 ft long by 9.8 ft wide and 3 ft high). The flush gate is 1.2 m (9.8 ft) long and 0.4 m (1.3 ft) high. It is stainless steel and is controlled hydraulically. The slope of the flush vault is 20%. The source of flushing water is local wastewater/combined sewer flows.

4.1.2.2 Type of wet weather flow treated

Combined sewer.

4.1.2.3 Equipment/vendor data

Not applicable.

4.1.2.4 Installation date

1988.

4.1.2.5 Construction cost

Not provided.

4.1.2.6 Frequency of use

Not provided.

4.1.2.7 Wet weather flow volume treated

Not provided.

4.1.2.8 Influent/effluent quality

Not applicable.

4.1.2.9 Method of monitoring

Not provided.

4.1.2.10 Operating history

Operators note fair performance in cleaning deposits during flushing.

4.1.2.11 Maintenance intensiveness

Operator indicated that there is a need for two persons to inspect the facility after each activation.

4.1.2.12 Annual O&M cost

Not provided.

4.1.3 Facility No. 3: Stadt Kirchhain, Germany

This facility has an off-line flushing gate system that is used to flush downstream sediments to a downstream regulator which then discharges into a 300 mm (12 in) sewer.

4.1.3.1 Design criteria

The system consists of two off-line CSO flushing vaults. Each facility has a pipeline which is 1600 mm (60 in), circular and 115 m (377 ft) in length with an average slope of 0.4%. Each facility has an off-line flushing vault holding 4 m^3 (1,057 gal) used to clean the pipe, and the flushing water source is local wastewater/combined sewer flows. The flush vault size is 2 m long by 2.5 m wide and 1 m high (6.6 ft long by 8.2 ft wide and 3.3 ft high). The flush gate is 1.2 m (3.9 ft) long and 0.4 m 1.3 ft high. It is stainless steel and is controlled electrically. The slope of the flush vault is 20%.

4.1.3.2 Type of wet weather flow treated

CSO.

4.1.3.3 Equipment/vendor data

Not applicable.

4.1.3.4 Installation date

1991.

4.1.3.5 Construction cost

Not provided.

4.1.3.6 Frequency of use

Not provided.

4.1.3.7 Wet weather flow volume treated

Not provided.

4.1.3.8 Influent/effluent quality

Not applicable.

4.1.3.9 Method of monitoring

Not provided.

4.1.3.10 Operating history

Operators note fair performance in cleaning deposits during flushing. This facility was modeled and the resulting terminal velocity of the sewer was calculated to be 0.6 m/s (2 ft/s). This value correlates with the reported performance by the operators.

4.1.3.11 Operational benefits and difficulties

None noted.

4.1.3.12 Maintenance intensiveness

Not provided.

4.1.3.13 Annual O&M cost

Not provided.

4.1.4 Facility No. 4: Stadt Heidenheim, Germany

This facility has an in-line flushing chamber located within the regulator structure on the dry weather sewer. Extreme high wet weather flows overtop the weir in the center of the structure into a 1200 mm (48 in) bypass conduit.

4.1.4.1 Design criteria

The facility has a pipeline which is 2200 mm (88 in), circular and 240 m (787 ft) in length. The average slope is not known. The in-line flushing vault holding 10 m^3 (2,642 gal) is used to clean the pipe, and the flushing water source is local wastewater/combined sewer flows. The flush vault size is 3 m long by 5 m wide and 0.7 m high (9.8 ft long by 16.4 ft wide and 2.3 ft high). The flush gate is 1.2 m (3.9 ft) long and 0.4 m (1.3 ft) high. It is stainless steel and is controlled hydraulically. The slope of the flush vault is 10%.

4.1.4.2 Type of wet weather flow treated

Combined sewer.

4.1.4.3 Equipment/vendor data

Not applicable.

4.1.4.3 Installation date

1993.

4.1.4.4 Construction cost

Not provided.

4.1.4.5 .Frequency of use

Not provided.

4.1.4.6 Wet weather flow volume treated

Not provided.

4.1.4.7 Influent/effluent quality

Not applicable.

4.1.4.8 Method of monitoring

Not provided.

4.1.4.9 Operating history

Operators note good performance in cleaning deposits during flushing. This facility was modeled, and the resulting terminal velocity of the sewer was calculated to be 1.0 m/s (3.3 ft/s). This value correlates with the reported performance by the operators.

4.1.4.10 Operational benefits and difficulties

None noted.

4.1.4.11 Maintenance intensiveness

There is a need for two persons to inspect the facility after each activation.

4.1.4.12 Annual O&M cost

Not provided.

4.1.5 Facility No. 5: Markt Grossostheim, Germany

This flushing vault and storage facility involves the use of an off-line storage compartment created by a spill weir from the dry weather sewer. Extreme high overflows can bypass the spill storage compartment. The bypasses are controlled by a mechanical operated bending weir. The downstream flow throttle is a mechanical knife valve controlled by an in-line magnetic flow meter.

4.1.5.1 Design criteria

The facility has a pipeline, which is 2200 mm (88 in), circular and 190 m (623 ft) in length with an average slope of 9.4%. The off-line flushing vault holding 15 m^3 is used to clean the pipe, and the flushing water source is local wastewater/combined sewer flows. The flush vault size is 6.5 m long by 2.2 m wide and 1.03 m high (21.3 ft long by 7.2 ft wide and 3.4 ft high). The flush gate is 1.5 m (4.9 ft) long and 0.4 m (1.3 ft) high. It is stainless steel and is controlled hydraulically. The slope of the flush vault is 15%.

4.1.5.2 Type of wet weather flow treated

Combined sewer.

4.1.5.3 Equipment/vendor data

Not provided.

4.1.5.4 Installation date

1993.

4.1.5.5 Construction cost

Not provided.

4.1.5.6 Frequency of use

Not provided.

4.1.5.7 Wet weather flow volume treated

Not provided.

4.1.5.8 Influent/effluent quality

Not applicable.

4.1.5.9 Method of monitoring

Not provided.

4.1.5.10 Operating history

Operators note excellent to good performance in cleaning deposits during flushing. This facility was modeled, and the resulting terminal velocity of the sewer was calculated to be 1.2 m/s (3.9 f/s). This value correlates with the reported performance by the operators.

4.1.5.11 Operational benefits and difficulties

None noted.

4.1.5.12 Maintenance intensiveness

There is a need for two persons to inspect the facility after each activation.

4.1.5.13 Annual O&M cost

Not provided.

4.1.6 Facility No. 6: Osterbruch-Opperhausen, Germany

This is an off-line flushwater chamber that is filled by a pumped river water source. The flushing gate facility is used to clean the downstream combined sewer.

4.1.6.1 Design criteria

The facility has a pipeline that is 250 mm (10 in), circular and 1000 m (3,281 ft) in length with an average slope of 0.1%. The off-line flushing vault holding 2 m^3 (528 gal) is used to clean the pipe and the flushing water source is river water. The flush vault size is 2.4 m long by 1.6 m wide and 0.75 m high (7.9 ft long by 5.2 ft wide and 2.5 ft high). The flush gate is 0.5 m (1.6 ft) long and 0.4 m (1.3 ft) high. It is stainless steel and is controlled hydraulically. The slope of the flush vault is 20%.

4.1.6.2 Type of wet weather flow treated

Combined sewer.

4.1.6.3 Equipment/vendor data

Not provided.

4.1.6.4 Installation date

1989.

4.1.6.5 Construction cost

Not provided.

4.1.6.6 Frequency of use

Not provided.

4.1.6.7 Wet weather flow volume treated

Not provided.

4.1.6.8 Influent/effluent quality

Not applicable.

4.1.6.9 Method of monitoring

Not provided.

4.1.6.10 Operating history

Operators note excellent to good performance in cleaning deposits during flushing.

4.1.6.11 Operational benefits and difficulties

None indicated.

4.1.6.12 Maintenance intensiveness

There is a need for two persons to inspect the facility after each activation.

4.1.6.13 Annual O&M cost

Not provided.

4.1.7 Facility No. 7: Geimeide Hettstadt, Germany

This is an off-line CSO pipeline that fills together with the flushing chamber during an overflow event.

4.1.7.1 Design criteria

The facility has a pipeline which is 1600 mm (64 in), circular and 224 m (735 ft) in length with an average slope of 0.5%. The off-line flushing vault holding 10 m^3 (2,642 gal) is used to clean the pipe, and the flushing water source is local wastewater/combined sewer flows. The flush vault size is 5 m long by 2.7 m wide and 0.75 m high (16.4 ft long by 8.9 ft wide and 2.5 ft high). The flush gate is 1.5 m (4.9 ft) long and 0.4 m (1.3 ft) high. It is stainless steel and is controlled hydraulically. The slope of the flush vault is 15%.

4.1.7.2 Type of wet weather flow treated

Combined sewer.

4.1.7.3 Equipment/vendor data

Not applicable.

4.1.7.4 Installation date

1992.

4.1.7.5 Construction cost

Not provided.

4.1.7.6 Frequency of use

Not provided.

4.1.7.7 Wet weather flow volume treated

Not provided.

4.1.7.8 Influent/effluent quality

Not applicable.

4.1.7.9 Method of monitoring

Not provided.

4.1.7.10 Operating history

Operators note good to fair performance in cleaning deposits during flushing.

4.1.7.11 Operational benefits and difficulties

None indicated.

4.1.7.12 Maintenance intensiveness

Operational protocol for this facility requires two persons to inspect the facility after each activation.

4.1.7.13 Annual O&M cost

Not provided.

4.2 Technology Evaluation

Pisano (1998) noted a strong correlation between operator-indicated performance and the modeled velocity of evaluated flushing systems. It seems prudent that before considering flushing systems for retrofit dynamic hydraulic analysis of flushing to ensure adequate velocities should be undertaken. Case studies indicate that velocities above 1 m/s (3.2 ft/s) are required for effective flushing. This may preclude many applications for retrofit in areas where adequate pipe slope is not available.

4.2.1 Applications

An overall ranking of flushing systems is provided in Table 4-1.

Table 4-1: Overall Ranking of Flushing Systems for Sewers

	Potential Applicability		
Application	**BOD/TSS Removal Performance**	**Life-Cycle Cost (Capital + O&M)**	**Notes**
Combined Sewer Overflows	High	Low	Best results are when terminal velocity of flushing wave is > 1 m/s
Sanitary Sewer Overflows	Not Applicable	Not Applicable	New regulations on SSOs prohibit discharges
Wet Weather Treatment at Treatment Facilities	Not Applicable	Not Applicable	Can be considered as a maintenance tool if large pipes are used only for CSOs
Urban Stormwater Runoff	High	Low	Best results are when terminal velocity of flushing wave is > 1 m/s

Detailed ranking of wet weather treatment performance factors for flushing systems are listed in Table 4-2. Each of these is discussed in more detail in the paragraphs that follow.

Table 4-2: Ranking of Detailed Performance Factors for Flushing Systems

Application	Performance Factor							
	Construction Cost	Annual O&M Cost	Startup Effort	Shutdown & cleanup effort	Treatment effectiveness	Implement-ability	Operational Intensiveness	Maintenance Intensiveness
CSO	L	L	L	L	M	L	L	L
SSO	NA	NA	NA	NA	NA	NA	NA	NA
Wet Weather Treatment at Treatment Facilities	L	L	L	L	L	L	L	L
Urban Stormwater Runoff	L	L	L	L	M	L	L	L
L = Low M = Medium H = High NA = Not Applicable								

4.2.2 Construction Cost

Construction/capital costs typical of flushing systems are presented in Table 4-3.

Table 4-3: Flushing Systems Typical Capital Costs

Alternative	Capital Cost $	Capital Cost $/m^2
Manual	10,000	2.91
Flushing Spray	680,000	197.67
Tipping Flusher	525,000	152.62
Flushing Gate	350,000	101.74

Source: Pisano et al (1998)

4.2.3 O&M Cost

Typical operation and maintenance costs for flushing systems are presented in Table 4-4.

Table 4-4: Flushing System Operation and Maintenance Costs

Alternative	O&M Cost $/ Event	O&M Cost $/m^2
Manual	6,600	1.92
Flushing Spray	1,548	0.45
Tipping Flusher	378	0.11
Flushing Gate	250	0.07

Source: Pisano et al (1998)

4.2.4 Startup Effort

From case studies reviewed, there is little startup effort required.

4.2.5 Shutdown and Cleanup Effort

From case studies reviewed, typically two people were allocated to visit the flushing systems after an occurrence to check on cleaning of system. No mention has been made concerning any unusual manpower needs for cleaning and maintenance.

4.2.6 Treatment Effectiveness

The goal of flushing sewers is to remove sediments in sewers and minimize the potential for hydrogen sulfide generation in the sewer.

4.2.7 Implementability

There has been more use of flushing gates for sewers in Europe than the United States. There needs to be further research in developing the relationships of minimizing sewer deposits within CSO and their impact on receiving waters.

4.2.8 Operational Intensiveness

From the case studies reviewed no operational attention is required with the use of flushing gates for sewers.

4.2.0 Maintenance Intensiveness

The flushing gate technology is very limited in the degree of maintenance required to keep the system operational.

4.3 Summary and Conclusions

In a comprchensive study of 18 applications of flushing technology, Pisano et al. (1998) found that the tipping flusher and flushing gate technology appear to be the most cost-effective means for flushing solids and debris from tanks. They also found the most efficient method for flushing large diameter flat depositing sewers is the flush gate technology.

The performance of both types of flushing equipment for tanks and flush gates for sewers has been rated as good to excellent. Based on calculations for most of the facilities using flushing gates, the terminal velocities at the end of the flushing wave exceeded 1 m/s (3.2 f/s). This terminal velocity was adequate for cleansing. Before consideration of retrofit into existing systems, dynamic hydraulic analysis of flushing wave velocities should be performed to insure that adequate velocities can be generated and solids removal achieved.

CHAPTER 5.0

FLOATABLE CONTROLS

There are several techniques available for the control of floatables in wet weather discharges (EPA, 1999d):

- Baffles - consist of vertical steel plates or concrete beams that extend from the top of the sewer to just below the top of the regulating weir. During an overflow event, the baffles retain floatables while water passes under the baffles, over the regulator, and into the receiving water body. When the flow recedes below the bottom of the baffle, floatable material is carried downstream to the wastewater treatment plant.
- Screens - consist of a series of vertical and horizontal bars or wires that trap floatables while allowing water to pass through the openings. Screens used for wet weather wastewater discharges control include mechanically cleaned permanent screens, static screens, traveling screens, or drum screens.
- Catch basin modifications – include technologies to prevent floatables from entering the catch basin and technologies that retain floatables within the catch basin. In each case, floatables must be gathered for disposal periodically.
- Netting – captured floatables through the use of mesh nets, either within the system or downstream of a discharge.
- Containment booms – containment systems that use specially fabricated floatation structures with suspended curtains designed to capture buoyant materials. They are typically anchored to a shoreline structure at the bottom, and they can be located downstream of one or more outfalls. After a storm event, material captured in the boom can be removed.
- Skimmer vessels - are specially designed boats used to collect floating debris, including material contained behind booms. Skimmer vessels almost always require companion equipment, including a shore conveyor for offloading, a truck for disposal, and a trailer for land transport.

Most floatable control programs utilize a combination of these controls to achieve the desired reduction of floatables (Leo et al., 1998), and they are most often applied to combined sewer overflows (CSOs).

The reader is referred to WERF report, *Assessment of Technologies for Screening, Floatable Control, and Screenings Handling* (stock no. 00CTS4), for a full assessment of these wet weather practices.

CHAPTER 6.0

OPERATIONAL ENHANCEMENTS

6.1 Literature Review

As discussed in Chapter 1, best practices include both technologies and methodologies (and combinations thereof) to achieve the required level of treatment for wet weather flows. This chapter focuses predominantly on the methodologies that can be used with existing wastewater systems to achieve the needed wet weather performance without expansion of existing works. Implementation of these methodologies may require the installation of new or additional technologies, but the primary focus is on improving the operational efficiency of the system.

In conducting this literature review, it is apparent that there has been a great increase in the last 10 years in the investigation and application of real time control (RTC) to improve efficiencies. RTC is a practice that encompasses a wide array of technologies and techniques that are often selected based on a systems analysis. Furthermore, the computer models generated during a systems analysis (or derivatives thereof) are often used to operate RTC practices. RTC may be applied in the collection system, at the treatment plant, or both. Beck et al. (1991) present an overview of the advantages and caveats of using a dynamic (as opposed to steady state) approach to the operation of collection systems and treatment plants.

6.2 Systems Analysis

The introduction of desktop computer simulation of wastewater collection and treatment systems has resulted in a significant increase in studies focused on improving the operational efficiency of the entire wastewater system (collection and treatment) and its interaction with the water environment. System analysis generally requires assessment of the systems over extended periods of time in order to adequately represent the performance of the system. This is commonly accomplished through long-term simulation of the processes involved and calibration and verification of the simulation results against measured performance.

Rauch and Harremoes (1996) describe the principles by which deterministic models of the wastewater collection system, the wastewater treatment works, and the receiving water can be linked to provide forecasts of water quality impacts under numerous wet weather scenarios. Commonly termed "integrated modeling", this approach is being increasingly applied in the European Union (FWR, 1994).

Vanderborght and Wollast (1990) discuss the continuous mentoring conducted within the City of Brussels, Belgium, to facilitate implementation and operation of RTC. Issues associated with selecting parameters for measurements, selecting monitoring equipment, and maintaining the monitoring network are presented.

Carstensen et al. (1996) discuss the issues involved with predictive control of wastewater systems and propose a hybrid model system to improve the accuracy and precision of control strategies. The hybrid models, termed "grey box" models, utilize deterministic models of the physical systems where they are well-established and then incorporating stochastic elements to improve correlation of model output to observed measurements. The authors view this approach as an emerging methodology and that models will likely need to be constructed on a case-by-case basis currently due to variations in systems and desired control applications. No performance data or design guidelines are provided.

Crabtree et al. (1996) described the application of integrated modeling for the City of Derby in the United Kingdom. The application was conducted under the Urban Pollution Management (UPM) Procedure (FWR, 1994) which recognizes integrated modeling as a viable approach to wet weather discharge management. The UPM Procedure resulted in a greater understanding of the wet weather phenomena experienced by the City and demonstrated that certain proposed system improvements (determined using other methods) would have no incremental environmental benefit. Revised modifications were under consideration at the time of publication and therefore no performance data or design guidelines are provided.

Gong et al. (1996) describe a neural network model to represent the management of a wet weather treatment facility in France. The neural network model was constructed to pull from both prior experience regarding the operational needs of the plant and from experimental data for training and validation. Rainfall was used as the parameter on which operational decisions were made. The neural network model was compared to deterministic models of the same system with good result when sufficient rain gage information was present. Limited performance data are provided.

Vanrolleghem et al. (1996) describe a decision support system methodology by which the cost impacts of treatment plant design and operation options can be assessed. This methodology includes attempts to address flexibility and robustness issues within analyses. These factors become especially relevant during wet weather flows. No performance data are provided.

6.3 Real-Time Control

It has long been recognized that high hydraulic loads can reduce treatment plant performance and even create additional wet weather pollution if treatment plant solids are "washed out" by high flows (Schnell, 1991, Kim and Pipes, 1996) requiring a extended period before the plant can resume adequate levels of biological treatment. Because of the risks associated with loss of biological solids in a facility, most plants are operated with significant factors of safety during wet weather to prevent this (Noland and Mather, 1981).

Beck and Lessard (1990) describe an effort to develop wet weather operational control strategies for treatment plants and couple those with real-time forecasting. By operating the plant immediately ahead of wet weather events to maximize plant capacity and minimize solids washout, the facility was able to maintain treatment capacity for smaller events that might have caused washout. Performance data are provided.

Nielsen et al. (1996) developed a new operational mode for treatment plants to go closer to critical process limits while balancing between hydraulic and biological capacities. The new mode involves Aeration Tank Settling (ATS) where air to aeration is turned off and on at the inset of peak wet weather flows for 2 hour periods. This allows much of the solids currently within the aeration basin to settle and not be carried over to secondary clarification where the risk of solids washout is high. The periods of aeration keep solids from becoming anaerobic. The concept was applied at a full-scale plant in Denmark and resulted in hydraulic capacity increase of at least 25% for periods of up to 2 weeks with no degradation in treatment plant performance. Performance data and design guidelines are provided.

Kim and Pipes (1996) describe an operational strategy similar to ATS in which aeration was turned off to an activated sludge process at a treatment facility in Chester, Pa. The benefits of this strategy were demonstrated on four wet weather events that typically would have caused washout of solids. Washout did not occur while treatment capacity was maintained. Performance data are provided.

Bertand-Krajewski et al. (1995) describe the continuous flow and pollutant measurements required to optimize treatment plant performance at a plant receiving flow from a combined sewer system north of Paris, France. The measurements are used to develop operational strategics to balance between storage and treatment within the plant. The additional data collected demonstrated both short-term and long-term effects on concentrations of biological and chemical oxygen demand, ammonia, and total suspended solids that dictate operational changes for the facility to improve efficiencies. No results from the operational changes are presented. Design guidelines on system monitoring can be derived from the study.

Muller and Krauth (1998) developed an automatic control strategy to allow treatment plants to go closer to critical process limits. The process utilized the ammonia concentration in the nitrification zone of an activated sludge system and the sludge blanket level in the secondary clarifier as the control parameters. The control strategy was evaluated at a full scale plant in Germany and demonstrated the ability of the plant to process 27% greater flow than previously processed. Limited performance data are provided. Design guidance is provided.

Nyberg et al. (1996) present the application of RTC to a treatment facility in Malmo, Sweden, to improve wet weather performance. The RTC system, developed through the use of deterministic hydraulic and treatment process models, involves the use of in-line storage in the sewer main leading to the plant, step feed operation within the plant, and, in extreme events, the by-pass of flows to avoid solids washout from clarifiers. Performance data are provided. Design guidelines are provided.

Durchschalg (1990) discusses the issues associated effectively performing long-term simulation of treatment works subject to wet weather flows and describes the results of one of the earlier efforts to simulate biochemical oxygen demand discharges from a treatment works subject to high flows using long-term models. Results of the simulations were used to develop operational strategies that were implemented off-line. Performance data are provided.

6.4 Impact of Wet Weather Flows on Operation of WWTPs

Although there is some commonality among different treatment plants, there are also differences that often reflect the broader differences in the geographic areas, the nature of the receiving waters, and the communities they serve. The objective function for operating a specific wastewater treatment plant could be based on one of the following goals:

- ***"Keep the plant running."*** The goal of the operator is to avoid gross process failure such as loss of solids or a serious disruption/deterioration of the biological activity.
- ***"Stay within the limits."*** In this situation, the operator wants to maintain the performance of the treatment plant within the performance boundaries established by the regulators. The goal is to keep the effluent quality from violating the standards.
- ***"Optimize the operation."*** The operation of the treatment can be adjusted to reduce the operating costs, most often in terms of reducing the use of energy or chemicals. The goal is to provide the adequate level of wastewater treatment at the lowest possible cost.

Operators generally try to meet the operational objectives listed above in the order of priority as shown. At a well-designed modern treatment plant, the objective of operation will normally include all of the objectives.

The scope of the WERF study (project no. 99-WWF-4, *Sensing and Control Systems: A Review of Municipal and Industrial Experiences,* stock no. D23101) includes a detailed look into issues related to process control in treatment plants. Additional information about automation and control of wastewater treatment plants can be found in the WEF Special Publication Automated Process Control Strategies (1997).

It is important to recognize that wet weather often introduces a significant disturbance into our system (treatment plant operation). During wet weather, significant increases in hydraulic loading can change the basic goals and the control objective function of our operation. Therefore, the process control strategies that were designed with the more "advanced" optimization goals are not necessarily relevant when the state of our system is experiencing a significant disturbance (i.e. during a heavy storm, the operator will seldom worry about optimizing the power consumption).

Operation of a wastewater treatment plant includes many types of processes: the two main categories are biological (including kinetics of degradation) and physical (movement of suspended solids and solids/liquid separation). During wet weather flow operation, the main stress is on the physical aspects of wastewater treatment: the hydraulics and the solids-liquid separation. Researchers and practitioners agree on one thing: during a storm, the secondary clarifiers are the most vulnerable unit in a wastewater treatment plant. The major danger to the WWTP most often observed during wet weather conditions is the loss of solids over the weirs of a secondary clarifier (Kim and Pipes, 1996; Nyberg et al., 1996; Nielsen et al., 1996; Rauch and Harremoes, 1996).

6.5 General Challenges to Process Control in WWTPs

The application of process control to wastewater treatment plants lags a bit behind the application of process control to other seemingly similar processes found in chemical engineering. Some of the challenges are listed below.

6.5.1 Instrumentation

A general lack of reliable and accurate instruments that can measure process parameters that are important and meaningful for the operation of the process. Some of the variables important to the wastewater treatment process are:

- Settling properties of sludge
- Characteristics of the influent wastewater
- Biological activity of the biomass
- Components/composition of the biomass

They are still not readily available from reliable on-line measuring units. While these variables are included (and computed) in the models, they are not easy to obtain in an on-line situation.

6.5.2 Process loads

Difficulty in predicting or controlling process loads. In most process industry, the input to the process (or a process unit) is known or controlled (or both). For most wastewater systems, several aspects of the input to the process (hydraulic load, magnitude and composition of the organic load) are difficult to predict.

6.5.3 Cost and volume

Very low cost of operation and very high volumes for processing. Even a small treatment plant has a higher throughput (in terms of volume) than a large refinery. The costs for processing wastewater are comparatively very low.

6.5.4 Funding

Funding cycles that favor capital improvements. In many cases, municipalities will receive grants for capital projects. Often, there is a desire to build as much capacity as possible and process control is still not as convincing to many administrators as adding extra capacity (with "brick and mortar" investments).

6.5.5 Understanding the Processes

Some key processes are still not well understood. This is especially true for our understanding of the dynamic behavior of clarifiers. Most analyses of settler operation today are conducted using oversimplified one-dimensional layered models (Vitasovic, 1989; Takacs et al., 1989) rather than more accurate descriptions of clarifier behavior (Vitasovic et al., 1997) that include the important effects of density currents.

6.5.6 Process control interactions

Analysis of process control is made more difficult because of important interactions between:

- Different process units in wastewater treatment plants. The operation design of specific process units impacts other process units; for example, the design and operation of the aeration basins impacts the secondary clarifier. The process units do not necessarily have to be directly connected physically in order for the interaction to occur; for example, the design of the influent pumping station may have an impact on the clarifiers as well.
- Design and operation. Sometimes the treatment plant design process does not fully take into account the operational issues. In those cases, the facilities may not have the flexibility required for implementing control strategies. Some design features that are important for control include, for example, the ability for step feed of the influent wastewater.
- Different components of an urban watershed. When one looks at an organizational chart for some municipalities, it might not be easy to deduce that sewer networks are actually physically connected to treatment plants. Yet, the operation and design of sewer networks can have a profound impact on the treatment plant. This interaction is especially strong during wet weather conditions: for example, every gallon of CSO reduction (in the sewer network) means a change to the inflow (and hydraulic load) to the treatment plant. Traditionally, these systems were treated separately, and it is only recently that a demand has intensified for an integrated approach.

Although these challenges apply to normal conditions (such as dry weather flow), they are exacerbated during transient events (such as storms) when the state of a system is shifting rapidly.

6.6 WWTP Operation Possibilities During Wet Weather Flows

During the intense transient events such as storms, the primary objective is most often to avoid experiencing a loss of solids from the system. The operator will normally have control over the process variables discussed below:

6.6.1 Control of the influent flow pattern distribution

Control of the influent flow pattern distribution among different biological reactors (step feed). While many activated sludge treatment plants have a single point where the influent wastewater is introduced into the biological reactor(s), the flexibility to distribute the wastewater to different points along the reactor provides the operator with valuable control options. The advantages of step feed were recognized fairly early (Torpey, 1948). Torpey shifted the influent from the head of the (full-scale) plant and introduced it at several points along the length of the reactor, thus creating a range of contacting patterns between the sewage and the mixed liquor. Sorensen (1980) examined the use of step feed control in a pilot plant in Denmark. Different contacting patterns were simulated and a control strategy based on Oxygen Uptake Rate (OUR) was examined with a goal to improve effluent quality, energy utilization, and excess sludge production. Buhr et al. (1984) looked into using the step feed to control the solids loading into the secondary clarifier. The work of Thompson et al. (1992) examines the impact of step feed control to reduce the impact of storm flows at the Woodward Avenue Water Pollution Control Plant in Hamilton Ontario. The objective function for the control strategy during normal operation was to ensure maximum nitrification, but during wet weather operators used a different control strategy to prevent loss of solids from the secondary clarifier.

6.6.2 Control of RAS flow rate

Control of the recycle activated sludge (RAS) flow rate includes the following typical strategies:

- Fixed RAS flow control
- Variable RAS control
- Blanket level control
- RAS flow control based on settled sludge method, and
- RAS flow control based on solids flux theory.

As RAS flow is increased, it is possible to speed up the withdrawal of solids from the clarifier and thus improve the stability of clarifier operation. However, the increase in the recycled flow will put more hydraulic pressure on the clarifier since this flow does return to the clarifier. Therefore, as we examine this control strategy we need to consider the status of other control actions, such as step feed control. A more in-depth summary of these control strategies can be found in the Water Environment Federation publication Automated Process Control Strategies (WEF, 1997).

6.6.3 Equalization of flow into the treatment plant

The system that delivers the flow to the treatment plant normally includes the sewer network and different types of storage facilities. This system includes volume that may be used for storage. It is possible to use the storage upstream of the treatment plant to equalize the flow to the treatment plant and reduce the hydraulic shocks to the system. Traditionally, the analysis of the treatment plant operation was often done independently from the analysis of the network dynamics. This approach assumes that the flows delivered to the plant cannot be controlled. Although sewer networks are obviously connected to treatment plants, traditional "way of doing business" often leads to an artificial (but real nonetheless) separation of these issues. For example, CSO reduction efforts will impact the flow to the treatment plant (and thus the plant operation) but in the past these two systems were studied and controlled separately. An integrated, holistic approach to the interaction between the sewer network and the treatment plant is increasingly proposed by engineers and regulators (FWR, 1994, Clifforde and Murrell, 1993, Rauch and Harremoes, 1996, Carstensen and Nielsen, 1996, Crabtree et al., 1996). The barriers to a more holistic approach might include the organizational ("turf") issues within agencies where these areas are often managed by different groups, and also a lack of truly integrated analysis and simulation tools.

An interesting approach has emerged recently as a way to alleviate the adverse impact of the hydraulic load to the secondary clarifiers during wet weather flows. A strategy called aeration tank settling (Nielsen et al., 1996, Kim and Pipes, 1996) includes shutting off the aeration mechanisms in the biological reactors during a storm event. During the few hours of the rain event, the aeration mechanisms are shut off, the solids settle down in the biological reactor, and thus the solids loading to the clarifier is reduced. Experience in full scale plants has shown encouraging results. However, caution is advised because the success of this unorthodox strategy may be impacted by many things including the design of the aeration mechanism (mechanical aerators or fine bubble diffusers), the impact on biological activity, and the impact on settling properties of sludge.

The variables listed above are those most commonly used to deal with the impact of wet weather on the treatment process. The best strategy will depend on a number of factors, and in some cases several different aspects are considered in a comprehensive control approach (Nyberg et al., 1996, Vanrolleghem et al., 1996).

6.7 Technology Evaluation

Many of the best practices that fall under the category of operational enhancements represent an emerging area of technology. The concept of improved operations to increase the efficiency of treatment works during wet weather is not new. What makes this an emerging area of technology has been the advances of the past decade in computing, instrumentation, and telecommunications.

6.7.1 Applications

Operational enhancements can be applied to the operation of collection systems to improve delivery of wet weather flows to treatment works. It can further be applied at the treatment works to ensure that flows are treated as desired and are not allowed to overwhelm the treatment process.

6.7.2 Construction Cost

Construction costs vary widely depending on application.

6.7.3 O&M Cost

Operation and maintenance costs are generally low but can be significant depending on the complexity of the system.

6.7.4 Startup Effort

The effort involved in startup for operational enhancements varies but is generally low and most enhancements use instrumentation to measure and control system components.

6.7.5 Shutdown and Cleanup Effort

Not applicable.

6.7.6 Treatment Effectiveness

Operational enhancements by definition are designed to ensure that existing treatment processes perform as desired.

6.7.7 Implementability

Highly variable. The value of operational enhancements versus other wet weather technologies needs to be evaluated on a case-by-case.

6.7.8 Operational Intensiveness

Since most operational enhancements rely on instrumentation and, in many cases, automation, operation intensiveness is generally low but can be significant.

6.7.9 Maintenance Intensiveness

Compared to other wet weather technologies, very low.

6.8 Summary and Conclusions

With all the challenges that are facing the operator, there is some encouraging news on the horizon.

Mathematical models are improving. Models for biological kinetics have been widely verified against real plant data, and even the clarifier models are significantly better today (Ekama et al., 1997, Vitasovic et al., 1997). Models capable of accurately predicting the behavior of secondary settling tanks exist today. These models are still proprietary and not yet widely available, but there are ongoing efforts to develop public-domain version of these models.

This will greatly improve our ability to explain and understand the operational and design issues related to clarifiers.

There is continuing progress in the field of instrumentation. Although progress has not been as swift as some of the practitioners had hoped for, the instrumentation vendors continue to make advances. A useful development in the area of instrumentation has been the work done by the Instrumentation Testing Association.

Operational improvements as a means to meet wet weather treatment goals should be considered on a case-by-case basis and should not only consider technologies such as computer modeling, instrumentation, and automation, but also the human element to which the technologies will be applied.

CHAPTER 7.0

DISINFECTION

7.1 Literature Review

Wastewater effluents intended for discharge to receiving waters required disinfection. This has typically been achieved with chlorine or ultraviolet radiation. There are numerous design texts (Metcalf & Eddy, 1991, WEF, 1991) that address the application of chlorination for disinfection and dechlorination for the removal of chlorine residuals. Heaney et al. (1996) and Field (1997) discuss the research needs associated with disinfection of wet weather wastewater flows. Chief among the research needs is the establishment of more reliable means of measuring and determining performance. Field et al. (1993) found that traditional indicator organisms are unreliable for health-risk measurement. They further discuss the impact of the variable nature of combined sewer flows (e.g., intermittency and variations in suspended solids and temperature) on design disinfection processes.

Ultraviolet (UV) radiation is the most common alternative used for disinfection after chlorine and chlorine derivatives (Metcalf and Eddy, 1991). Tchobanoglous (1999) describes the factors affecting UV disinfection efficiency and conclude that particles in wastewater have a predominate effect, an important consideration for the application of UV to wet weather flows.

Darby et al. (1991) found in a survey of increasing application of UV at municipal wastewater plants and conducted a comparison between UV and chlorination that included a model to guide application. UV presents a number of benefits over chlorine including much fewer safety concerns and elimination of the need to remove disinfectant residuals before discharge. UV also has a relatively fast startup time, generally requires less space, and could be housed at remote locations within the wastewater system. These attributes make it applicable to periodic wet weather flows. Conversely, the large capital investment and long residence times required to achieve organism inactivation work against UV for use on infrequent wet weather flows. Energy costs also weigh against UV. Darby et al. (1991) present a model for predicting performance of UV systems and compare the economic feasibility of UV to that of chlorination.

Laine et al. (1998) designed a storm water treatment plant that combines air flotation, filtration, and UV disinfection. The system has been tested at the pilot plant level in the eastern suburbs of Paris, France. The UV disinfection process has resulted in the microbiological contamination of the effluent being reduced to a level in compliance with the European Union regulations that define the quality of bathing water.

Maney (1996) describes an automated wastewater treatment facility used to treat periodic wet weather flows in DuPage County, IL. The system, a separate wet weather flow treatment facility, can be activated remotely, is intended to be operated by one person, and uses chlorine for disinfection. Performance data are not presented. Some design guidelines are presented.

Gayle and Mitchell (1996) described the design and implementation of satellite treatment facilities for wet weather flows in Deerfield, IL. The treatment facilities use rotating fine screens and chlorination to achieve a maximum discharge concentration of 30 mg/L of BOD and TSS with no more than 0.75 mg/L of residual chlorine. Dechlorination was not required. Some performance data are presented, as are some design guidelines.

Sorber (1983) discussed the removal of viruses from wastewater and effluent by various treatment processes. Included in his assessment were the following disinfection processes: chlorination, ozone, and chlorine dioxide. UV was not assessed. Performance data are presented; design guidelines are not.

Boner et al. (1995) describe a bench-scale test of UV disinfection in conjunction with a modified vortex separator. Disinfection efficiency was a function of separator solids removal performance. Full-scale testing was recommended. Performance data are presented.

Boner and Turner (1996) present the application of full-scale testing of UV disinfection in conjunction with a modified vortex separator in Columbus, GA. Results of the project have not been reported.

Arnett and Boner (1998) describe a range of chemical disinfection techniques being tested in Columbus, GA, on wet weather flows from a vortex separator. The techniques include sodium hypochlorite, chlorine dioxide, and peracetic acid. Dechlorination was accomplished with sodium bisulfite. Performance data are provided.

FWR (1994) discusses the disinfection alternatives for consideration by wastewater systems considering wet weather flow treatment in order to meet United Kingdom and European Union water quality standards for combined sewer discharges. Disinfectants presented include: UV, ozone, peroxyacetic acid, and chlorine. Ozone is presented as effective but too costly. No performance data or design guidelines are provided.

7.2 Performance Data

The following information is provided from operating facilities which screen and disinfect as part of their CSO Control Plan. This study found little in the way of the application of UV for the disinfection of wet weather flows. UV can be particularly effective in a relatively controlled environment and has none of the environmental concerns that chlorine poses. However, UV is particularly sensitive to the amount of particles (suspended solids) in the stream to be disinfected, and much variation in the amount of particles can profoundly affect the transmissivity (and disinfecting capability) of UV. Furthermore, UV requires substantial capital cost and its operating costs are high because of the power required.

As such, UV systems have not been seen as widely applicable to wet weather flows outside of the controlled environment of the treatment plant. The facilities cited in this report are using chlorine (in some form) for disinfection, although some investigated other means.

7.2.1 Facility No. 1: Spring Creek Auxiliary Water Pollution Control Plant, New York City, N.Y.

The main objective of this pilot study was to demonstrate alternatives to hypochlorite disinfection for application at this facility. The disinfection technologies investigated included UV irradiation, ozonation, chlorine dioxide disinfection, chlorination/dechlorination, and electron beam irradiation (E-beam).

7.2.1.1 Design criteria

The Spring Creek AWPCP is a CSO detention facility with a minimum of 50,375 m^3 (13.3 million gallon) storage capacity in the basin and 11,625 m^3 (3.1 million gallon storage) in the influent conduit. Currently, liquid sodium hypochlorite is used for disinfection. It is manually controlled, mixed with carrier water and injected into the influent conduits. The dosage is controlled to maintain a residual of 1.5 mg/L.

7.2.1.2 Type of wet weather flow treated

Combined sewer.

7.2.1.3 Equipment/vendor data

Not provided.

7.2.1.4 Installation date

Not provided.

7.2.1.5 Construction cost

Chlorination/dechlorination and chlorine dioxide were determined to be the most cost-effective technologies; the capital costs are shown below.

Capital Cost (K $)		Flow	
Chlorine Dioxide	Chlorination / Dechlorination	m^3/d	mgd
695	912	3,059,000	808
1,159	1,045	6,116,000	1,615
1,931	1,219	12,234,000	3,230

The annual costs for ozone were 7 to 10 times the annual costs of these technologies based on peak design rates of 3,059,000 m^3/d (808 mgd), 6,116,000 m^3/d (1,615 mgd) and 12,234,000 m^3/d (3,230 mgd), while the annual costs for UV were 14 to 27 times the annual costs of the two technologies for the same peak flow rates.

7.2.1.6 Frequency of use

Not provided.

7.2.1.7 Wet weather flow volume treated

Not provided.

7.2.1.8 Influent/effluent quality

Not provided.

7.2.1.9 Method of monitoring

Not provided.

7.2.1.10 Operating history

Generally, all of the tested disinfection technologies, with the exception of E-beam, were able to effectively provide bacterial reductions of three to four logs. Chlorination/dechlorination, chlorine dioxide, and ozone were able to provide the required levels of disinfection over the entire range of wastewater quality tested. UV disinfection tended to decrease at higher TSS levels (TSS > than 150 mg/L). To achieve a four-log reduction of fecal coliform and fecal coliform effleunt concentrations of less than 1,000 colony forming units/100 mL (cfu/100 mL), required dosages each of the technologies was:

UV = 60–80 mWs/cm^2
O_3 = 24 mg/L
ClO_2 = 8–10 mg/L
Cl_2 = 20–28 mg/L.

7.2.1.11 Operational benefits and difficulties

Because an effective chlorine, gas-free process of chlorine dioxide generation has not been proven reliable and because chlorine gas cannot be transported within New York City, disinfection with chlorine dioxide could not be recommended. Only chlorination/dechlorination using sodium hypochlorite and sodium bisulfate was recommended.

7.2.1.12 Maintenance intensiveness

Not provided.

7.2.1.13 Annual O&M cost

Not provided.

7.2.2 Facility No. 2: Conner Creek Pilot CSO Control Facility, Detroit, Mich.

The Conner Creek Pilot CSO Control Facility is intended to disinfect, settle, and skim CSO and is sized based on 5 minutes of detention time at the 10-year, 1-hour peak flow of 32,447,000 m^3/d (8,567 mgd). The resulting retention basin volume is 116,250 m^3 (30 million gallons). The retention basin is preceded by screening and high-rate disinfection utilizing sodium hypochlorite.

7.2.2.1 Design criteria

Data from a disinfection field testing program demonstrated that the daily fecal coliform limit of 400 colony forming units (cfu) per 100 mL was always achieved with a CT of 100 or higher. The design of the NaOCl feed system was based on providing a CT of 125 with all units in service at the 10 year, 1 hr peak flow of 32,446,656 m^3/d (8,751 mgd). The NaOCl feed pumps are required to provide a maximum chlorine dose of 25 mg/L to meet the NPDES permit requirements for effluent fecal coliforms.

7.2.2.2 Type of wet weather flow treated

Combined sewer.

7.2.2.3 Equipment/vendor data

The aspects of NaOCl storage based on disinfecting 6,316, 250 m^3 (1,630 million gallons) of CSO and assuming an average chlorine dosage of 20 mg/L include the following:

- Required chlorine to disinfect the overflow volume is 123,423 kg (271,856 lb);
- Average daily CSO volume of 37,588 m^3 (9.7 million gallons) yields 735 kg (1,618 lb);

Total volume turnover is 168 days. A degradation analysis was completed based on receiving 15% NaOCl solution and diluting it to 10%, which decreases the decomposition rate of the chemical. Assuming daily withdrawal and refilling, the NaOCl will degrade to a strength of 7.54% in 168 days.

Based on the ability to deliver only three trucks per day of NaOCl at 15% strength and each truck with a capacity of 18,600 L (4,800 gallons) provides a total NaOCl storage of 1,473 m^3 (380,000 gallons). To store this requires four rubber-lined steel tanks each with a volume of 368 m^3 (95,000 gallons). The tanks are located outside in a chemical storage area that provides containment for 130% of the volume of one tank or 478 m^3 (123,500 gallons).

7.2.2.3.1 NaOCl feed system

The NaOCl feed system is designed for the 10 yr, 1 hr peak flow with 7.54% NaOcl to provide a CT of 125 and a chlorine dose of 25 mg/L. This provides a feed rate of 124 L/s (1,970 gpm). With one pump out of service, the system can still deliver a CT of 112.5 or a chlorine dose of 22.5 mg/L for the design storm flow which still exceeds the minimum CT of 100. The range of flows to be treated vary from 1,223,000 m^3/d (323 mgd) to 32,430,000 m^3/d (8,567 mgd). The NaOCl feed system requires a turndown of 27:1. To achieve this turndown requirement, the feed system is configured with 10 magnetically driven centrifugal pumps, one per screen, with each pump having a feed line and a recirculation line each with motorized diaphragm valves. Each pump is sized to provide a maximum capacity of 12.41 L/s (197 gpm) for a total feed capacity of 124.10 L/s (1,970 gpm). The feed rate is monitored by a magnetic flow meter.

During dry weather flow, and when the retention basin is not operating, the NaOCl pumps will operate sequentially, recirculating 12.41 L/s (197 gpm) to the tanks. This will both mix the tanks and flush the pumps and piping of salt deposits. Each of the 10 pumps will operate for 3 hours each day, with a different pump running each day for 10 calendar days. At day 11, the process resets itself.

7.2.2.3.2 Mixing

Three different mixers were tested at this facility and were:

- US Filter Stranco Products, 14.9 kW (20 Hp) Water Champ (induction/vacuum unit)
- Master Company, 14.9 kW (20 Hp) Gas Master (induction/vacuum unit)
- Master Company, 14.9 kW (20) Hp Gas Master (non-induction/vacuum unit)

These mixers showed similar results with respect to mixing effectiveness. During testing for a "uniform mix," dye was at the mixer which indicated a rising plume phenomenon which was attributed to the vacuum induction of the mixers decreasing the density of the feed. As a result, it was recommended that each of the screening/mixing channels be provided with 14.9 kW (20 HP) mixers of the non-vacuum type.

There are a total of six mixers at this facility, each rated at 14.9 kW (20 Hp).

7.2.2.4 Miscellaneous Data

Not provided were the installation date, construction cost, frequency of use, wet weather flow volume treated, influent/effluent quality, method of monitoring, operating history, operational benefits and difficulties, maintenance intensiveness, and annual O&M cost.

7.2.3 Facility No. 3: Redford, Mich.

This CSO Treatment facility is operated by the Wayne County Department of Environment which provides for storing and treating overflows on the Rouge River in metropolitan Detroit. This particular facility was designed to handle flows from a combined drainage area of 2.7 km^2 (669 acres). Its design storm is a one year-one hour storm. The basin provides for screening, storage, skimming, and chlorination.

The Redford facility has its screen located in the wet well prior to the pumps. Bar spacing is 1.91 cm (0.75 in). Skimming is accomplished by a 161 m (528 ft) weir/baffle prior to the effluent weir. The total surface area of the basins is 2,210 m^2 (23,760 sq ft), and the design flow rate is 465,000 m^3/d (123 mgd). The surface overflow rate is 211.36 $m^3/d/m^2$ (5,180 gpd/sq ft).

Chlorination is achieved by using liquid sodium hypochlorite with the feed point at the discharge of the influent pumps.

7.2.3.1 Design criteria

The basin configuration is two parallel compartments preceded by one vortex separator with a diameter of 10.67 m (35 ft). The basin dimensions are 54.86 m by 20.12 m by 3.41 m (180 ft by 66 ft by 11.2 ft) in each compartment. The volume of each compartment is 368.13 m^3 (0.9 million gallons) for a total volume of 7,363 m^3 (1.9 million gallons). The detention time is 20 minutes. The bacteriological design criteria required that the disinfection system provide for a NPDES effluent limit for fecal coliform of 400 per 100 mL. An effluent goal for total residual chlorine (TRC) was set at 1 mg/L.

7.2.3.2 Type of wet weather flow treated

Combined sewer.

7.2.3.3 Equipment/vendor data

Sodium hypochlorite is used for disinfection. The storage concentration used in the design criteria was 5%, at a design feed rate of 10 mg/L. The resulting hypochlorite storage tank volume was two 45.4 m^3 (12,000 gal) fiberglass tanks, based on two back-to-back design storms. There is a secondary containment area in the basement. An air mixing system was incorporated into the design of the tanks which is not used by the operators. The chemical feed pumps are Pulsafeeder, double diaphragm, variable speed. The feed rate for the hypochlorite is adjusted by the number of pumps operating. The chemical feed piping is Schedule 80 PVC with PVC ball or diaphragm valves. The ability to flush the chemical piping with clean water is provided.

7.2.3.4 Installation date

1997.

7.2.3.5 Construction cost

Not available.

7.2.3.6 Frequency of use

Between June 1997 and March 2000, the Redford facility annual overflow frequency was average of 2.4 times during the period from November to April and 3.2 times during the period from May to October.

7.2.3.7 Wet weather flow volume treated

The following data represent the effluent volume discharged from the CSO facility for recorded events from 1997 through 1999.

		Total Volume (million gallons)	
Date	**Rainfall (in)**	**Influent**	**Effluent**
16 Jun 97	0.63	2.29	0.10
2 Jul 97	1.75	11.45	8.58
26 Jul 97	0.82	2.43	0.16
10 Sep 97	1.71	3.54	0.99
7/8 Jan 98	0.21	2.82	0.79
17/21 Feb 98	2.42	25.85	23.42
9/10 Mar 98	1.05	6.43	3.48
9/10 Apr 98	1.21	3.57	0.90
6 Aug 98	3.42	8.76	5.92
7 Oct 98	1.86	2.35	0.42
22 Jan 99	1.06	2.95	0.23
22/26 Apr 98	1.23	11.45	4.86
12 Jun 99	2.46	3.30	2.07
27 Jun 99	2.01	3.29	0.5
28/29 Jun 99	0.70	1.75	0.06

7.2.3.8 Influent/effluent quality

The average treated effluent quality for these overflow events was the following:

Date	Average Treated Effluent Concentration (mg/L)			
	CBOD5	TSS	NH3	Total P
16 Jun 97	NO SAMPLES			
2 Jul 97	21.4	189	4.60	0.85
26 Jul 97	NO SAMPLES			
10 Sep 97	20.4	54	2.02	1.21
7/8 Jan 98	19.6	24	4.70	0.67
17/21 Feb 98	26.3	93	1.83	0.88
9/10 Mar 98	14.2	75	2.69	0.76
9/10 Apr 98	17.5	71	2.4	0.95
6 Aug 98	13.5	110	1.01	1.09
7 Oct 98	23.2	123	2.2	5.92
22 Jan 99	4.5	145	3.5	1.75
22/26 Apr 98	14.2	69	2.94	1.08
12 Jun 99	NO SAMPLES			
27 Jun 99	34	61	2.5	1.15
28/29 Jun 99	NO SAMPLES			

7.2.3.9 Method of monitoring

The monitoring program was initially designed to provide information required to meet NPDES requirements. In subsequent meetings with the state agencies, the sampling program was modified to demonstrate whether a "first flush" occurred which had an effect on the quality of the CSO basin influent. The goals of the program were:

- Quantification of all loads captured in the basin
- Quantification of all effluent loads from the basin
- Intermediate point sampling for multi-cell basins
- Pollutant concentration variability data
- Flow and volume data

Within 6 months of the start of the sampling effort, parts of the protocol were changed to reflect experience gained. For example, effluent samples were not collected at 15 and 30 minute intervals for CSO basins with influent pump stations, the sampling interval was changed to coincide with the collection of a target of three samples per pump cycle. During the evaluation period, samples were required every 15 minutes for the first hour, then every 30 minutes for the next 2 hours. In addition to discrete samples, composite samples were required for the influent and effluent flows. Grab samples were also required every hour for fecal coliform, dissolved oxygen, temperature, and pH.

7.2.3.10 Operating history

The facility has been in operation since 1997.

7.2.3.11 Operational benefits and difficulties

Overall, two years of experience shows that this facility and others which are part of the Wayne County CSO evaluation program can be operated to achieve the required bacteriological standard of fecal coliforms not to exceed 400 per 100 mL and meet a TRC of less than or equal to 2 mg/L at the end of the pipe. A more sophisticated or flexible chlorine feed and control system may be required. There has also been reported the tendency for the operators to overrule the hypochlorite pacing system and manually control the dosage. Other reported issues are the actual strength of the hypochlorite when used since it degrades over time and the under recording of flow rates at low flows by the multipath transit time flow meters.

7.2.3.12 Maintenance intensiveness

The majority of the maintenance issues appear to be directed to the need to constantly repair leaks associated with the hypochlorite piping.

7.2.3.13 Annual O&M cost

The annual operating costs for the Redford CSO Facility are:

Labor - \$107,900
Energy and Utilities – Not available
Chemicals and Supplies – \$22,600
Laboratory and Services - \$25,000
Total - \$155,000

7.2.4 Facility No. 4: Inkster, Mich.

This CSO treatment facility is operated by the Wayne County Department of Environment which provides for storing and treating overflows on the Rouge River in metropolitan Detroit. This particular facility was designed to handle flows from a combined drainage area of 2.084 km^2 (515 acres). Its design storm is a one year-one hour storm. The basin provides for screening, storage, skimming and chlorination.

The Inkster facility has its screen located in the wet well prior to the pumps. Bar spacing is 1.91 cm (0.75 in). Skimming is accomplished by a 144 m (472 ft) weir/baffle prior to the effluent weir. The total surface area of the basins is 2,073 m^2 (22,300 sq ft), and the design flow rate is 558,144 m^3/d (147 mgd). The surface overflow rate is 269.3 m^3/d/m^2 (4.6 gpm/sq ft).

Chlorination is achieved by using liquid sodium hypochlorite with the feed point at the discharge of the influent pumps.

7.2.4.1 Design criteria

The basin configuration is two parallel detention compartments preceded by one first flush tank with a volume of 4,263 m^3 (1.1 million gallons). The parallel detention basin dimensions are 54.86 m by 18.29 m by 3.58 m (180 ft by 60 ft by 11.75 ft) in each compartment. The volume of each compartment is 3,875 m^3 (1 million gallons) for a total volume of 12,013 m^3 (3.1 million gallons) including the first flush tank. The detention time is 20 minutes. The bacteriological design criteria required that the disinfection system provide for a NPDES effluent limit for fecal coliform of 400 per 100 mL. An effluent goal for total residual chlorine (TRC) was set at 1 mg/L.

7.2.4.2 Type of wet weather flow treated

Combined sewer.

7.2.4.3 Equipment/vendor data

See above narrative for Redford, MI, CSO Facility.

7.2.4.4 Installation date

1997.

7.2.4.5 Construction cost

Not applicable.

7.2.4.6 Frequency of use

Between June 1997 and March 2000, the Inkster facility annual overflow frequency was an average of 3.2 times during the period from November to April and 3.9 times during the period from May to October.

7.2.4.7 Wet weather flow volume treated

The following data represent the effluent volume discharged from the CSO facility for recorded events from 1997 through 1999.

		Total Volume (million gallons)	
Date	**Rainfall (in)**	**Influent**	**Effluent**
21 Jun 97	0.73	6.42	2.64
2 Jul 97	1.75	5.05	2.10
10 Sep 97	1.51	6.93	2.88
7/8 Jan 98	5.62	1.73	14.5
17/20 Feb 98	2.56	44.04	40.00
9/10 Mar 98	1.23	17.86	13.71
9/10 Apr 98	1.32	9.44	4.92
26 Apr 98	1.40	6.32	1.02
31 May 98	1.07	4.16	0.06
7/8 Jul 98	1.11	4.65	0.43
6 Aug 98	2.18	8.23	4.29
22/24 Jan 99	1.20	14.35	1.73
22/24 Apr 99	1.06	10.82	3.99
12/13 Jun 99	2.14	7.52	3.72
13/14 Jun 99	0.97	5.18	1.96
24/25 Jun 99	0.99	4.24	0.89
27/30 Jun 99	3.98	24.66	13.89
23/24 Jul 99	1.89	6.92	2.07

7.2.4.8 Influent/effluent quality

The average treated effluent quality for these overflow events was the following:

	Average Treated Effluent Concentration (mg/L)			
Date	**CBOD5**	**TSS**	**NH3**	**Total P**
21 Jun 97	25.3	152	3.9	0.76
2 Jul 97	12.8	164	2.8	1.1
10 Sep 97	<10	71	0.22	0.51
7/8 Jan 98	14.5	55	0.82	0.58
17/20 Feb 98	18.4	144.8	0.53	0.66
9/10 Mar 98	4.5	103.6	0.22	0.58
9/10 Apr 98	25.4	99.4	0.86	0.69
26 Apr 98	30	50	0.78	0.58
31 May 98	NO SAMPLES			
7/8 Jul 98	NO SAMPLES			
6 Aug 98	22.6	117	0.36	0.65
22/24 Jan 99	17.5	108	1.35	1.41
22/24 Apr 99	15.2	65.8	1.44	0.87
12/13 Jun 99	NO SAMPLES			
13/14 Jun 99	NO SAMPLES			
24/25 Jun 99	NO SAMPLES			
27/30 Jun 99	21.8	125	0.66	0.79
23/24 Jul 99	29	151	0.14	5.7

7.2.4.9 Method of monitoring

See above narrative for Redford, Mich., CSO facility.

7.2.4.10 Operating history

The facility has been in operation since 1997.

7.2.4.11 Operational benefits and difficulties

See above narrative for Redford, Mich., CSO facility.

7.2.4.12 Maintenance intensiveness

See above narrative for Redford, Mich., CSO facility.

7.2.4.13 Annual O&M cost

The annual operating costs for the Inkster CSO facility are:

Labor - $133,000
Energy and Utilities – $85,300
Chemicals and Supplies – $23,100
Laboratory and Services - $20,000
Total - $261,400

7.2.5 Facility No. 5: Dearborn Heights, Mich.

This CSO treatment facility is operated by the Wayne County Department of Environment, which provides for storing and treating overflows on the Rouge River in metropolitan Detroit. This particular facility was designed to handle flows from a combined drainage area of 1.46 km^2 (360 acres). Its design storm is a ten year-one hour storm. The basin provides for screening, storage, skimming and chlorination.

The Dearborn Heights facility has its screen located in the wet well prior to the pumps. Bar spacing is 1.91 cm (0.75 in). Skimming is accomplished by a 137 m (450 ft) weir/baffle prior to the effluent weir. The total surface area of the basins in 2,930 m^2 (31,500 sq ft) and the design flow rate is 499 m^3/d (132 mgd). The surface overflow rate is 171.4 $m^3/d/m^2$ (4,200 gpd/sq ft).

Chlorination is achieved by using liquid sodium hypochlorite with the feed point at the discharge of the influent pumps.

7.2.5.1 Design criteria

The basin configuration is three parallel detention compartments with the capability to use the first tank as a "first flush tank". The compartment dimensions are 53.3 m by 18.3 m by 3.54 m (175 ft by 60 ft by 11.6 ft) in each compartment. The volume of each compartment is

3,488 m^3 (0.9 million gallons) for a total volume of 10,463 m^3 (2.7 million gallons). The detention time is 30 minutes. The bacteriological design criteria required that the disinfection system provide for a NPDES effluent limit for fecal coliform of 400 per 100 mL. An effluent goal for total residual chlorine (TRC) was set at 1 mg/L.

7.2.5.2 Type of wet weather flow treated

Combined sewer.

7.2.5.3 Equipment/vendor data

See above narrative for Redford, MI, CSO Facility.

7.2.5.4 Installation date

1977.

7.2.5.5 Construction cost

Not provided.

7.2.5.6 Frequency of use

Between June 1998 and March 2000, the Dearborn Heights facility annual overflow frequency was an average of 2.1 times during the period from November to April and 1.7 times during the period from May to October.

7.2.5.7 Wet weather flow volume treated

The following data represent the effluent volume discharged from the CSO facility for recorded events from 1998 through 1999.

		Total Volume (million gallons)	
Date	**Rainfall (in)**	**Influent**	**Effluent**
8 Jan 1998	0.23	7.47	3.84
17/20 Feb 1998	2.66	33.61	30.21
9/10 Mar 1998	1.11	10.63	7.05
9/10 Apr 1998	1.41	5.04	1.57
6/7 Aug 1998	2.20	12.00	7.00
22/25 Apr 1999	1.33	17.53	4.70
12/13 Jun 1999	2.98	7.67	3.93
20/29 Jun 1999	2.75	12.14	9.84
23/24 Jul 1999	1.93	5.79	1.80

7.2.5.8 Influent/effluent quality

The average treated effluent quality for these overflow events was the following:

	Average Treated Effluent Concentration (mg/L)			
Date	CBOD5	TSS	NH3	Total P
8 Jan 1998	43.2	71.7	4.47	1.01
17/20 Feb 1998	20.6	71.1	2.22	0.68
9/10 Mar 1998	12.6	48	2.33	1.26
9/10 Apr 1998	NO SAMPLES			
6/7 Aug 1998	23.8	82.7	2.21	0.66
22/25 Apr 1999	4.6	53.3	2.26	0.92
12/13 Jun 1999	5.7	86.25	1.15	1.13
20/29 Jun 1999	5.29	44.4	0.26	0.57
23/24 Jul 1999	NO SAMPLES			

7.2.5.9 Method of monitoring

See above narrative for Redford, Mich., CSO facility.

7.2.5.10 Operating history

The facility has been in operation since 1997.

7.2.5.11 Operational benefits and difficulties

See above narrative for Redford, Mich., CSO facility.

7.2.5.12 Maintenance intensiveness

See above narrative for Redford, Mich., CSO facility.

7.2.5.13 Annual O&M cost

The annual operating costs for the Dearborn Heights CSO facility are:

Labor - $132,900
Energy and Utilities – $71,600
Chemicals and Supplies – $22,000
Laboratory and Services - $15,000
Total - $241,500

7.2.6 Facility No. 6: Acacia Park, Mich.

This CSO Treatment facility is operated by the Oakland County Drain Commissioner, which provides for storing and treating overflows on the Rouge River in metropolitan Detroit. This particular facility was designed to handle flows from a combined drainage area of 3.3 km^2 (816 acres). Its design storm is a one year-one hour storm. The basin provides for screening, storage, skimming, and chlorination.

The Acacia Park facility has its screen located in the effluent. Bar spacing is 1.91 cm (0.75 in). Skimming is accomplished by 592 m (1,942 ft) of weir/baffle located at the influent weir, before the intermediate weir, after the intermediate weir and before the effluent weir troughs. The total surface area of the basins is 2,380 m^2 (25,600 sq ft), and the design flow rate is 709,000 m^3/d (187 mgd). The surface overflow rate is 306 $m^3/d/m^2$ (7,500 gpd/sq ft).

Chlorination is achieved by using liquid sodium hypochlorite.

7.2.6.1 Design criteria

The basin configuration is two compartments in series. The compartment dimensions are 48.8 m by 24.4 m by 6.1 m (160 ft by 80 ft by 20 ft) in each compartment. The total volume of the compartments is 15,140 m^3 (4 million gallons). The detention time is 30 minutes. The bacteriological design criteria required that the disinfection system provide for a NPDES effluent limit for fecal coliform of 400 per 100 mL. An effluent goal for TRC was set at 1 mg/L.

7.2.6.2 Type of wet weather flow treated

Combined sewer.

7.2.6.3 Equipment/vendor data

Sodium hypochlorite is used for disinfection. The storage concentration used in the design criteria was 6–8%, at a design feed rate of 11.6 mg/L. The resulting hypochlorite storage tank volume was two 38.75 m^3 (10,000 gal) fiberglass tanks. There is a secondary containment area in the basement. Mixing is provided by pumping from tank to tank. The chemical feed pumps are Gorman Rupp, air diaphragm. There are three pumps provided for each basin. The feed rate for the hypochlorite is adjusted by flow controllers for the pump air strokes (two 7.62 cm (3 in)). The chemical feed piping is PVC with PVC valves. The ability to flush the chemical piping with clean water requires that a hose be added.

7.2.6.4 Installation date

1997.

7.2.6.5 Construction cost

Not provided.

7.2.6.6 Frequency of use

Between June 1998 and March 2000, the Acacia Park facility annual overflow frequency was an average of 2.0 times during the period from November to April and 2.5 times during the period from May to October.

7.2.6.7 Wet weather flow volume treated

Not provided.

7.2.6.8 Influent/effluent quality

The typical effluent quality at the Acacia Park CSO facility is as follows:

CBOD5	10 mg/L
TSS	29 mg/L
Ammonia	0.04 mg/L
Phosphorous	0.15 mg/L

7.2.6.9 Method of monitoring

See above narrative for Redford, Mich., CSO facility.

7.2.6.10 Operating history

The facility has been in operation since 1997.

7.2.6.11 Operational benefits and difficulties

See above narrative for Redford, Mich., CSO facility. In addition, the diaphragm pumps were observed by operators to be too large. Essentially, the pumps would operate at one stroke per three minutes at low flows.

7.2.6.12 Maintenance intensiveness

See above narrative for Redford, MI CSO Facility.

7.2.6.13 Annual O&M cost

The annual operating costs for the Acacia Park CSO facility are:

Labor - $58,600
Energy and Utilities – $24,800
Chemicals and Supplies – $25,900
Laboratory and Services - $10,500
Total - $119,800

7.2.7 Facility No. 7: Bloomfield Village, Mich.

This CSO treatment facility is operated by the Oakland County Drain Commissioner, which provides for storing and treating overflows on the Rouge River in metropolitan Detroit. This particular facility was designed to handle flows from a combined drainage area of 7.0 km^2 (1,735) acres. Its design storm is a one year-one hour storm. The basin provides for screening, storage, skimming, and chlorination.

The Bloomfield Village facility has its screen located in the effluent. Bar spacing is 1.91 cm (0.75 in). Skimming is accomplished by 760 m (2,495 ft) of weir/baffle located at the influent weir, before the intermediate weir, after the intermediate weir and before the effluent weir troughs. The total surface area of the basins is 6,601 m^2 (70,980 sq ft) and the design flow rate is 1,712,000 m^3/d (452 mgd). The surface overflow rate is 276 $m^3/d/m^2$ (6,760 gpd/sq ft). Chlorination is achieved by using liquid sodium hypochlorite.

7.2.7.1 Design criteria

The basin configuration is three compartments filling in series through different elevation weirs. The compartment dimensions are 48.0 m by 39.2 m by 6.1 m (157.5 ft by 128.5 ft by 20 ft) in each compartment. The total volume of the compartments is 38,750 m^3 (10 million gallons). The detention time is 30 minutes. The bacteriological design criteria required that the disinfection system provide for a NPDES effluent limit for fecal coliform of 400 per 100 mL. An effluent goal for TRC was set at 1 mg/L.

7.2.7.2 Type of wet weather flow treated

Combined sewer.

7.2.7.3 Equipment/vendor data

See above narrative for Acacia Park facility. There are three inflow controllers for the pump air strokes at this facility.

7.2.7.4 Installation date

1997.

7.2.7.5 Construction cost

Not provided.

7.2.7.6 Frequency of use

Between June 1998 and March 2000, the Acacia Park facility annual overflow frequency was an average of 1.4 times during the period from November to April and 1.0 times during the period from May to October.

7.2.7.7 Wet weather flow volume treated

Not provided.

7.2.7.8 Influent/effluent quality

Not provided.

7.2.7.9 Method of monitoring

See above narrative for Redford, Mich., CSO facility.

7.2.7.10 Operating history

The facility has been in operation since 1997.

7.2.7.11 Operational benefits and difficulties

See above narrative for Redford, Mich., CSO facility. In addition, the diaphragm pumps were observed by operators to be too large. Essentially, the pumps would operate at one stroke per three minutes at low flows.

7.2.7.12 Maintenance intensiveness

See above narrative for Redford, Mich., CSO facility.

7.2.7.13 Annual O&M cost

The annual operating costs for the Bloomfield Village CSO facility are:

Labor - $133,600
Energy and Utilities – $42,100
Chemicals and Supplies – $61,000
Laboratory and Services - $24,200
Total - $260,900

7.3 Technology Evaluation

7.3.1 Applications

An overall ranking of disinfection technologies are provided in Table 7–1 and 7–2.

Table 7–1: Overall Ranking of Chemical Disinfection for Various Applications

Application	Potential Applicability Fecal Coliform Reduction	Life-Cycle Cost (Capital + O&M)	Notes
CSO	High	Medium	Highly dependent on flow pacing
SSO	High	Medium	Highly dependent on flow pacing
WWT at Treatment Facilities	High	Medium	Highly dependent on flow pacing
Urban SW Runoff	High	Medium	Highly dependent on flow pacing

WWT, wet weather treatment; SW, stormwater

Table 7–2: Overall Ranking of Ultraviolet Disinfection for Various Applications

Application	Potential Applicability Fecal Coliform Reduction	Life-Cycle Cost (Capital + O&M)	Notes
CSO	Low	High	Highly dependent on water with minimal solids
SSO	Low	High	Highly dependent on water with minimal solids
WWT at Treatment Facilities	Low	High	Highly dependent on water with minimal solids
Urban SW Runoff	Low	High	Highly dependent on water with minimal solids

WWT, wet weather treatment; SW, stormwater

Detailed ranking of wet weather treatment performance factors for disinfection technologies are listed in Table 7–3 and 7–4. Each of these is discussed in more detail in the paragraphs that follow.

Table 7–3: Ranking of Detailed Performance Factors for Chemical Disinfection

Application	Performance Factor Cost		Effort				Intensiveness	
	Construction	Annual O&M	Startup	Shutdown & cleanup	Treatment effectiveness	Implement-ability	Operational	Maintenance
CSO	L	M	L	M	H	L	L	M
SSO	L	M	L	M	H	L	L	M
WWT at Treatment Facilities	L	M	L	M	H	L	L	M
Urban SW Runoff	L	M	L	M	H	L	L	M

WWT, wet weather treatment; SW, stormwater

Table 7–4: Ranking of Detailed Performance Factors for Hypochlorite Disinfection

Application	Performance Factor							
	Cost		Effort				Intensiveness	
	Construct-ion	Annual O&M	Startup Effort	Shutdown & cleanup	Treatment effectiveness	Implement-ability	Operational	Maintenance
CSO	L	M	L	M	H	L	L	M
SSO	L	M	L	M	H	L	L	M
WWT at Treatment Facilities	L	M	L	M	H	L	L	M
Urban SW Runoff	L	M	L	M	H	L	L	M

WWT, wet weather treatment; SW, stormwater

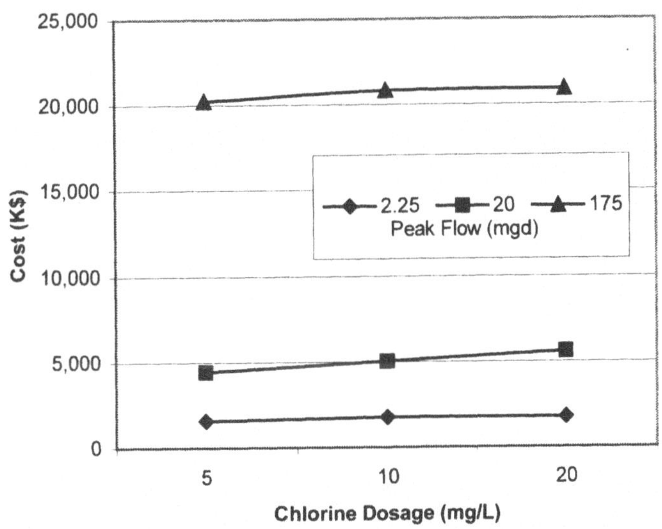

Figure 7–1: Capital Cost for Chlorination/Dechlorination

7.3.2 Construction Cost

Typical construction costs for chlorination/dechlorination systems are shown in Figure 7-1 (Darby, 1995, adjusted for inflation per ENR Construction Cost Indices).

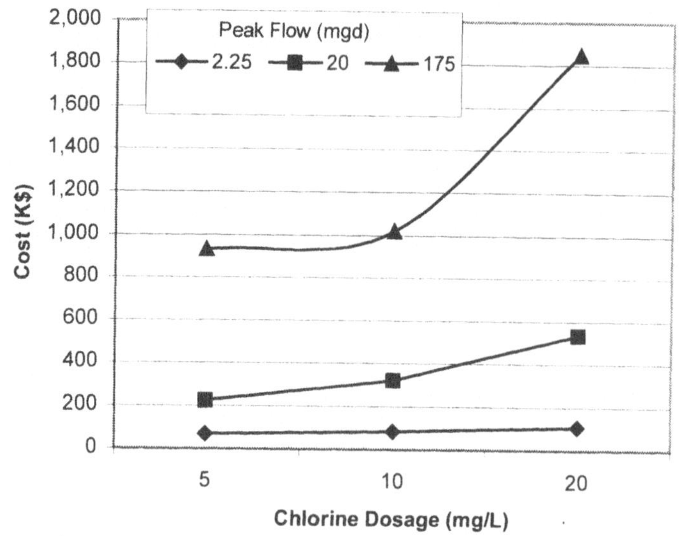

Figure 7–2: Operation and Maintenance Costs for Chlorination/Dechlorination

7.3.3 O&M Cost

Typical operation and maintenance costs for chlorination/dechlorination systems are shown in Figure 7–2 (Darby, 1995, adjusted for inflation per ENR Construction Cost Indices).

7.3.4 Startup Effort

In using hypochlorite, the stored chemical must be checked for its strength. The shelf life for hypochlorite is limited.

7.3.5 Shutdown and Cleanup Effort

Flushing water needs to be provided for chemical lines taken out of service. Hypochlorite is a strong corrosive, and gaskets on piping and piping joints will tend to fail. A good preventative maintenance program needs to be implemented.

7.3.6 Treatment Effectiveness

The treatment effectiveness in the use of disinfection is for fecal coliform reduction only. BOD and TSS removals should not be expected.

7.3.7 Implementability

Chlorination/dechlorination can generally be easily implemented at treatment facilities. Remote location of chlorination/dechlorination facilities can present several challenges, not the least of which are safety and security concerns of locating chemicals in areas of the system where they might not be attended constantly.

7.3.8 Operational Intensiveness

Operational intensiveness is generally low when systems are properly maintained and well instrumented.

7.3.9 Maintenance Intensiveness

The use of hypochlorite will require preventive maintenance procedures to be implemented to maintain the piping systems.

7.4 Summary and Conclusions

A variety of disinfection technologies are available to treat wet weather wastewater flows. Experience to date has shown that, considering a combination of performance, reliability, and cost, chlorination/dechlorination is generally the best choice for treating intermittent wet weather flows. The use of hypochlorite and sodium bisulfite is becoming more prevalent for chlorination/dechlorination as safety issues about the use of gaseous chlorine and sulfur dioxide are addressed. The effluent limits for chlorine and the bacteriological requirements for fecal coliforms determine the need for dechlorination. As more stringent fecal coliform limits are required which dictate a higher dosage rate of chlorine, the effluent may have a higher chlorine residual than allowed which then requires the use of a dechlorination process to conform to the effluent limitations.

The quantity of suspended solids in the treated CSO effluent impacts the efficiency of the disinfection process. As the level of suspended solids increase, the efficiencies of the disinfection process decrease and increased dosage rates of the disinfectant must be used to achieve the required bacteriological requirements. With the use of chlorine or hypochlorite, this can be achieved by increasing the dosage rate of the chemical. The use of UV and other technologies require that these processes must be sized for a higher peak capacity than normal flow conditions which result in higher capital costs for these alternative technologies. UV also requires substantial energy to operate which also provides concerns for operating costs.

UV and other technologies appear less viable for intermittent wet weather treatment, but could have opportunities for application. UV and other technologies may be excellent choices when considering the startup and shutdown of the disinfection facilities for a CSO application and when considering the security aspects of maintaining chemical storage for disinfection. However, the high capital and potentially high energy costs of these systems coupled with the intermittent use of the facilities often precludes the use of UV and other technologies.

REFERENCES

Alvarez, C.; J.A. Juanes; J.A. Revilla; K. Koev; A. Roldan and V. Ivanov (1999), "Environmental Study of the Alternatives for the Sewer System of a Small Coastal Community in the Bay of BISCAY," Water Science Technology, Vol. 39, Elsevier Science Ltd.

Anon (1995), "Combined Sewer Overflows: Guidance for Screening and Ranking," Government Reports Announcements & Index (GRA&I), No. 13.

Arnett, C. and M. Bonner (1999), "Columbus, Georgia Advance Demonstration Facility for Wet Weather Treatment Technologies," Proceedings of the 5th EPA National Wastewater Treatment Technology Transfer Workshop, EPA Document EPA 907/R-98/001, U.S. Environmental Protection Agency, Office of Research and Development, Washington, DC.

Beck, M. and P. Lessard (1990), "Operational Water Quality Management: Control of Storm Sewage at a Wastewater Treatment Plant," Research Journal WPCF, Vol. 62, No. 6, Water Environment Federation, Alexandria, Va.

Beck, M., J. Adeloye, B. Finney, and P. Lessard (1991), "Operational Water Quality Management: Transient Events and Seasonal Variability," Water Science and Technology, Vol. 24, No. 6, International Association for Water Quality (IAWQ), United Kingdom.

Bertrand-Krjaewski, J.-.L, M., Lefebvre, B. Lefai, and J.-M. Audic (1995), "Flow and Pollutant Measurements in a Combined Sewer System to Operate a Wastewater Treatment Plant and Its Storage Tank During Storm Events," Water Science and Technology, Vol. 31, No. 5, International Association for Water Quality (IAWQ), United Kingdom.

Boner, M. and B. Turner (1996), "Demonstration Testing of Multiple Wet-Weather Treatment Technologies – Columbus, Georgia," Conference Proceedings: Urban Wet Weather Pollution, Controlling Sewer Overflows and Stormwater Runoff, Water Environment Federation, Alexandria, VA.

Boner, M.C.; D.R. Ghosh; S.R. Harper; B.G. Turner (1995), "Modified Vortex Separator And UV Disinfection For Combined Sewer Overflow Treatment," Water Science and Technology, Vol. 31, No. 3-4, International Association for Water Quality (IAWQ), United Kingdom.

Booker, N., G. Ocal, and A. Preistley (1996), "Novel High-Rate Processes for Sewer Overflow Treatment," Water Science and Technology, Vol. 34, No. 3-4, International Association for Water Quality (IAWQ), United Kingdom.

Bridoux, G., A. Villeroux, M. Riotte, and M. Huau (1998), "Optimised Lamella Settling Process for Run Off Water Treatment," Water Science and Technology, Vol. 38, No. 10, International Association for Water Quality (IAWQ), United Kingdom.

Brombach, H.; C. Xanthopoulos; H.H. Hahn and W.C. Pisano (1993), "Experience with Vortex Separators for Combined Sewer Overflow Control," Water Science and Technology WSTED4, Vol. 27, No. 5-6.

Burgess, E.H.; L.A. Roesner; M.M. Umberg and T.A. Saygers (1991), "Planning Subsurface Storage-Transport Facilities For Combined Sewer Overflow Control In Cincinnati, Ohio," Proceeding the 1991 Specialty Conference on Environmental Engineering, Reno, Nev, USA.

Carstensen, J., M. Nielsen and P. Harremoes (1996), "Predictive Control of Sewer Systems By Means of Grey-Box Models," Water Science and Technology, Vol. 34, No. 3-4, International Association for Water Quality (IAWQ), United Kingdom.

Chandler, R.W. and W. Lewis (1977), "Control Of Sewer Overflows By Polymer Injection," Report EPA-600/2-77-189, NTIS No. PB 272 654, U.S. Environmental Protection Agency.

Columbus Water Works (2001), "Draft Summary Report on Wet Weather Demonstration Projects: CSO Technology Testing, Source Water Assessment and Protection, and Watershed Assessment and Management," prepared by WWETCO, LLC, available from Columbus Water Works, http://www.cwwga.org.

Crabtree, B., W. Earp, and P. Whalley (1996) "A Demonstration of the Benefits of Integrated Wastewater Planning for Controlling Transient Pollution," Water Science and Technology, Vol. 33, No. 2, International Association for Water Quality (IAWQ), United Kingdom.

Daligault, A., D. Meaudre, D. Arnault, V. Duc, N. Bardin, N. Aires, D. Biau, J. Schmid, P. Clement, and J.-Y. Viau (1991), "Stormwater and Lamella Settlers: Efficiency and Reliability," Water Science and Technology, Vol. 24, No. 6, International Association for Water Quality (IAWQ), United Kingdom.

Darby J., M. Heath, J. Jacangelo, F. Loge, P. Swaim, and G. Tchobanoglous (1995), "Comparison of UV Radiation to Chlorination: Guidance for Achieving Optimal Performance," Water Environment Research Foundation Report, Project 91-WWD-1, Water Environment Research Foundation, Alexandria, VaA.

Delft (2000), Department of Sanitary Engineering, Delft University of Technology - Faculty of Civil Engineering and Geosciences - The Netherlands, Internet: http://www.ct.tudelft.nl/wmg/sanitary/stowares.htm).

Ding, Y., Dresnack, R., and Chan, P. (1999) "Assessment of High-Rate Sedimentation Processes: Microcarrier Weighted Coagulation Jar Tests," EPA Internal Report Document EPA 600/x-99-033, U.S. Environmental Protection Agency, Office of Research and Development, Cincinnati, Ohio.

Durchschlag, A. (1990) "Long-Term Simulation of Pollutant Loads in Treatment Plant Effluents and Combined Sewer Overflows," Water Science and Technology, Vol. 22, No. 10-11, International Association for Water Quality (IAWQ), United Kingdom.

Dziopak, J. and J. Niemczynowicz (1999), "Vacuum-Driven CSO Detention Tanks," Urban Water, Vol. 1, No. 1.

EPA (1976), "The Swirl Concentrator for Erosion Runoff Treatment," EPA Document EPA-600/2-76-271, U.S. Environmental Protection Agency, Office of Research and Development, Cincinnati, Ohio.

EPA (1981), "Joint Dry/Wet Weather Treatment of Municipal Wastewater at Clatskanie, Oregon," EPA Document EPA 600/2-81-061, U.S. Environmental Protection Agency, Office of Research and Development, Washington, DC.

EPA (1982), "Swirl and Helical Bend Pollution Control Devices," EPA Document EPA 600/8-82-013, U.S. Environmental Protection Agency, Office of Research and Development, Cincinnati, Ohio.

EPA (1992), "Evaluation of Wet Weather Design Standards for Controlling Pollution from Combined Sewer Overflows," EPA Document EPA 230/R-92/006, U.S. Environmental Protection Agency, Office of Policy Analysis, Washington, DC.

EPA (1996), "Risk Management Research Plan for Wet Weather Flows," U.S. Environmental Protection Agency, Office of Wastewater Management, Washington, DC, Internet: http://www.epa.gov/owm/wwfrp1.htm.

EPA (1996a), "Assessment of Vortex Solids Separators for the Control and Treatment of Wet Weather Flow," Washington, DC, EPA 832-B-96-006, October 1996.

EPA (1998a), "Sewer and Tank Sediment Flushing: Case Studies," EPA Document EPA-600/R-98-157, U.S. Environmental Protection Agency, Office of Research and Development, Washington, DC.

EPA (1998b), "Storage/Sedimentation Facilities for Control of Storm and Combined Sewer Overflows: Design Manual," EPA Document EPA 600/R-98/006, U.S. Environmental Protection Agency, Office of Research and Development, Cincinnati, Ohio.

EPA (1999a), " Interim Report: Development of Bench-Scale Settling Apparatus: Settling Velocity Data for Design and Operation of Wet-Weather Flow Solids-Liquid Separation Processes," EPA Document EPA/600/x-99/031, U.S. Environmental Protection Agency, Office of Research and Development, Washington, DC.

EPA (1999b), " Internal Report: Assessment of High-Rate Sedimentation Processes: Microcarrier Weighted Coagulation Jar Tests," EPA Document EPA/600/x-99/033, U.S. Environmental Protection Agency, Office of Research and Development, Washington, DC.

EPA (1999c), "Innovative Urban Wet Weather Flow Management Systems," EPA Document EPA 600/R-99/029, U.S. Environmental Protection Agency, Office of Research and Development, Cincinnati, Ohio.

EPA (1999d), "Combined Sewer Overflow Technology Fact Sheet: Floatable Control," EPA Document EPA 832/F-99/008, U.S. Environmental Protection Agency, Office of Water, Washington, DC.

EPA (2000), "Wet Weather," U.S. Environmental Protection Agency, Office of Wastewater Management, Washington, DC, Internet: http://www.epa.gov/owm/wet.htm.

EPA (2000b), " Research Report: Retrofitting Control Facilities For Wet-Weather Flow Treatment," EPA Document EPA/600/R-00/020, U.S. Environmental Protection Agency, Office of Research and Development, Washington, DC.

EPA Region 6 (1995), "EPA Region 6 Water Management Division's Strategy for Wet Weather Sanitary Sewer Overflows," http://www.epa.gov/Region6/6en/w/sso/ssodraft.txt.

Field R. and Lager, J. (1975), "Urban Runoff Pollution Control -- State-of-the-Art," Journal of the Environmental Engineering Division, February 1975, American Society of Civil Engineers, New York, NY.

Field, R. (1986), "State-of-the-Art Update on Combined Sewer Overflow Control," Critical Reviews in Environmental Control Vol. 16, No. 2.

Field, R. (1997), "Storm and Combined Sewer Overflow: An Overview of EPA's Research Program," U.S. Environmental Protection Agency, Office of Research and Development, Cincinnati, Ohio, Internet: http://www.epa.gov/ednnrmrl/repository/epa88905.htm.

Field, R., Averill, D., O'Connor, T. and Steel, P. (1997), "Vortex Separation Technology," Water Quality Research Journal Canada, Vol. 32, No. 1, Canadian Association for Water Quality (CAWQ), Canada.

Field, R., M. O'Shea, and M. Brown (1993), " The Detection and Disinfection of Pathogens In Storm-Generated Flows," Water Science and Technology, Vol. 28, No. 3-5, International Association for Water Quality (IAWQ), United Kingdom.

Field, R.; R. Pitt; D. Jager and M. Brown (1995a), "Combined Sewer Overflow Control through In-Receiving Water Storage: An Efficiency Evaluation," Government Reports Announcements & Index (GRA&I), No. 5.

Field, R.; R. Pitt; M. Brown and T. O'Connor (1995b), "Combined sewer overflow control using storage in seawater," Water Research, Vol. 29, No. 6.

Freedman, D.S. and M.C. Jordan (1995), "Application of CSO-related bypasses in Maine," Proceeding 22nd Annual Conference on Integrated Water Resources Planning for the 21st Century, American Society of Civil Engineers, New York, NY.

Fuhs, G. (2000), "Stringent Bypass Rules Upheld by Federal Court," in Watershed & Wet Weather Technical Bulletin, Vol. 5, No. 3, July 2000, Water Environment Federation, Alexandria, VA.

FWR (1994), "Urban Pollution Management: A Planning Guide for the Management of Urban Wastewater Discharges During Wet Weather," Foundation for Water Research, Buckinghamshire, United Kingdom.

Gall, B., D. Averill, and D. Weatherbee (1997), "Modelling, Design, and Optimization of a Combined Sewer Overflow Treatment System," Water Quality Research Journal Canada, Vol. 32, No. 1, Canadian Association for Water Quality (CAWQ), Canada.

Gayle, D. and D. Mitchell (1996), "Innovative and Economical Sanitary Sewer Overflow Treatment Using Fine Screens and Chlorination," in Seminar Publication: National Conference in Sanitary Sewer Overflows, EPA Document EPA-625/R-96/007, U.S. Environmental Protection Agency, Office of Research and Development, Washington, DC.

Gong, N., X. Ding, T. Denoeux, J.-L. Betrand-Krajewski, and M. Clement (1996) "A Demonstration of the Benefits of Integrated Wastewater Planning for Controlling Transient Pollution," Water Science and Technology, Vol. 33, No. 2, International Association for Water Quality (IAWQ), United Kingdom.

Grey, G. and F. Oliveri (1999), "The Role of Catch Basins in a CSO Floatables Control Program," Proceedings of WETEC 1999, Water Environment Federation, Alexandria, Va.

H.I.L. Technology (2001), "Downstream Defender," Proprietary technology fact sheet, H.I.L. Technology, Inc., http://www.hil-tech.com/prdhom.html.

Heaney, J., L. Wright and D. Sample (1998), "Wet Weather Flow Management – A Research Needs Survey for Urban Areas," Water Environment Research Foundation Report, Project 96-IRM-1, Water Environment Research Foundation, Alexandria, Va.

Heaney, J., R. Field and D. Sullivan (1979), "Nationwide Cost of Wet Weather Pollution Control," Journal WPCF, Vol. 51, No. 9, Water Pollution Control Federation (now Water Environment Federation), Alexandria, Va.

Heinzmann, B. (1994) "Coagulation and Flocculation of Stormwater from a Separate Sewer System – A New Possibility for Enhanced Treatment," Water Science and Technology, Vol. 29, No. 12, International Association for Water Quality (IAWQ), United Kingdom.

Holtzman S.A. (1994), "Cyanide and Heavy Metal Removal," Advanced Chemical Technology, LLC, Internet: http://www.actglobal.net/wastewat.htm.

Holzer, P. and P. Krebs (1998), "Modelling the total ammonia impact of CSO and WWTP effluent on the receiving water," Water Science and Technology, Vol. 38, No. 10.

Huebner, M. and W. Gieger (1996), "Characterisation of the Performance of an Off Line Storage Tank," Water Science and Technology, Vol. 34, No. 3-4, International Association for Water Quality (IAWQ), United Kingdom.

Hvitved-Jacobsen, T. and K. Schaarup-Jensen (1990), "Dissolved Oxygen Stream Model for Combined Sewer Overflows," Water Science and Technology WSTED4, Vol. 22, No. 10/11.

Infilco Degremont Inc, "IS Accelator® Clarifier," P.O. Box 71390, Richmond, Va 23255-1390 Internet: http://www.infilcodegremont.com/NSACCEL.HTM.

Infilco Degremont Inc, "NS Accelator® Clarifier," P.O. Box 71390 • Richmond, VA 23255-1390 Internet: http://www.infilcodegremont.com/ISACCEL.HTM.

Ji, Z., Z. Vitasovic, and S. Zhou (1996), "A Fast Hydraulic Numerical Model for Large Sewer Collection Systems," Water Science and Technology, Vol. 34, No. 3-4, International Association for Water Quality (IAWQ), United Kingdom.

Johnson, C., Igwe, T., Mitchell, D., and Kaunelis, V. (2000), "Operating Experience with Large CSO Control Facilities," Proceedings of WEFTEC 2000, Water Environment Federation, Alexandria, VA.

Kaufman, H.L., and F. Lai (1980), "Review of Alternatives For Evaluation Of Sewer Flushing Dorchester area – Boston," Report EPA-600/2-80-118, NTIS No. PB 81-1 648, U. S. Environmental Protection Agency.

Kim, Y. and W. Pipes (1996), "Solids Routing in an Activated Sludge Process During Hydraulic Overloads," Water Science and Technology, Vol. 34, No. 3-4, International Association for Water Quality (IAWQ), United Kingdom.

Krebs, P.; P. Holzer; J.L. Huisman and W. Rauch (1999), "First flush of dissolved compounds," Water Science and Technology, Vol. 39, No. 9.

Krüger, Inc. (1998), "ACTIFLO® Provides Fast, Effective Treatment Of CSOs, Municipal Wastewater," Krüger FLUENTlines, Summer, Vol. 3 No. 1, Internet: http://krugerworld.com/USA/Fluentlines/vol3no1.html.

Krüger, Inc., "ACTIFLO® Clarification Microsand Enhanced Flocculation and Settling," Internet: http://Krugerworld.com/USA/actiflo.html.

Laine, S., T. Poujol, S. Dufay, J. Baron, and P. Robert (1998), "Treatment of Storm Water to Bathing Water Quality By Dissolved Air Flotation, Filtration, and Ultraviolet Disinfection," Water Science and Technology, Vol. 38, No. 10, International Association for Water Quality (IAWQ), United Kingdom.

Leo, W., W. Kurtz, and R. Gaffoglio (1998), "Comprehensive Planning for Control of CSO Floatables and Settleable Solids in New York City," Proceedings of WEFTEC 1999, Water Environment Federation, Alexandria, Va.

Lessard, P. and M. Beck (1990), "Operational Water Quality Management: Control of Storm Sewage at a Wastewater Treatment Plant," Journal WPCF, Vol. 62, No. 6, Water Pollution Control Federation (now Water Environment Federation), Alexandria, VaA.

Lorenzen, A., Ristenpart, E., and Pfuhl, W. (1996), "Flush Cleaning of Sewers," Water Science and Technology, Vol. 33, No. 9, International Association for Water Quality (IAWQ), United Kingdom.

Maney, S. (1996), " Automated Treatment Plant Reduces Sanitary Sewer Overflows," in Seminar Publication: National Conference in Sanitary Sewer Overflows, EPA Document EPA-

625/R-96/007, U.S. Environmental Protection Agency, Office of Research and Development, Washington, DC.

McCormick, E., J. Hegdal and R. Fischer (1992), "Dual-Purpose Treatment Plant Protects San Francisco Bay," Public Works, February, 1992, American Public Works Association, Washington, DC.

McGhee, T., "Water Supply and Sewerage," Sixth Edition, McGraw-Hill, Inc., New York, NY.

Medina, M., W. Huber, and J. Heaney (1981), "Modeling Stormwater Storage/Treatment Transients: Applications," Journal of the Environmental Engineering Division, Vol. 104, No. 7, American Society of Civil Engineers, New York, NY.

Metcalf & Eddy (1991), "Wastewater Engineering: Treatment, Disposal, and Reuse," Third Edition, McGraw-Hill, Inc., New York, NY.

Michelbach, S. and C. Wohrle (1993), "Settable Solids in a Combined Sewer System: Settling Characteristics – Pollution Load – Storm Water Tanks," Water Science and Technology, Vol. 27, No. 12, International Association for Water Quality (IAWQ), United Kingdom.

Muller, J. and K. Krauth (1998), "Wastewater Flow Management to Maximise the Capacity of Sewage Treatment Plants," Water Science and Technology, Vol. 37, No. 9, International Association for Water Quality (IAWQ), United Kingdom.

Nielsen, M., J. Carstensen, and P. Harremoes (1996), "Combined Control of Sewer and Treatment Plant During Rainstorm," Water Science and Technology, Vol. 34, No. 3-4, International Association for Water Quality (IAWQ), United Kingdom.

Noland, R. and W. Mather (1981), "Design to Optimize Multi-Stage Unite Processes For Peak Flows," Journal WPCF, Vol. 53, No. 3, Water Pollution Control Federation (now Water Environment Federation), Alexandria, VA.

NSF, Generic Verification Protocol for Vortex Separation – Draft 4.2, NSF International, http://www.nsf.org/etv/wwf/stormwater.html#list, 2001.

Nyberg, U., B. Andersson, and H. Aspergren (1996) "Real Time Control for Minimizing Effluent Concentrations During Strom Water Events," Water Science and Technology, Vol. 34, No. 3-4, International Association for Water Quality (IAWQ), United Kingdom.

O'Connor, T.; R. Field, D. Fischer, R. Rovansek, R. Pitt, S. Clark, and M. Lama (1999), "Urban Wet Weather Flows," Water Environment Research, Vol. 71, No. 5, Water Environment Federation, Alexandria, Va.

Pisano, W. , and C. Queriroz (1977), "Procedures For Estimating Dry Weather Pollutant Deposition In Sewerage Systems," Report EPA-600/2-77-120, NTIS No. PB 270 695, U. S. Environmental Protection Agency.

Pisano, W. et al. (1979), "Dry-Weather Deposition And Flushing For Combined Sewer Overflow Pollution Control," Report EPA-600/2-79-133, NTIS No. PB 80-118 524, U. S. Environmental Protection Agency.

Pisano, W. (1993), "Summary: The Fluidsep™ Vortex Solids Separator Technology," WK Inc., Marketing Brief, Belmont , Mass.

Pisano, W. Barsanti, J., Joyce, J., and Sorenson, H., (1998), "Sewer and Tank Sediment Flushing: Case Studies," Report EPA-600/R-98-157, NTIS No. PB 99-127839, U. S. Environmental Protection Agency.

Plum, V.; C.P. Dahl; L. Bentsen; C.R. Petersen; L. Napstjert and N.B. Thomsen (1998), "The Actiflo method," Water Science and Technology, Vol. 37, No. 1.

Rauch, W. and P. Harremoes (1996), "The Importance of the Treatment Plant Performance During Rain to Acute Water Pollution," Water Science and Technology, Vol. 34, No. 3-4, International Association for Water Quality (IAWQ), United Kingdom.

Reemtsma, T. R. Gnirb, and M. Jekel (2000), "Infiltration Of Combined Sewer Overflow And Tertiary Municipal Wastewater: An Integrated Laboratory And Field Study On Nutrients And Dissolved Organics," Water Research, Vol. 34, No. 4.

Roesner, L. and R. Carrier (1996), "Design and Operation of a Wet Weather Storage Facility to Eliminate Sanitary Sewer Overflows in the Vicinity of a Publicly-Owned Treatment Works Using an Extended-Period Simulation Model," in Seminar Publication: National Conference in Sanitary Sewer Overflows, EPA Document EPA-625/R-96/007, U.S. Environmental Protection Agency, Office of Research and Development, Washington, DC.

Roesner, L.A. and E.H. Burgess (1992), "The Role of Computer Modeling in Combined Sewer Overflow Abatement Planning," Water Science and Technology WSTED4, Vol. 26, No. 7-8.

Roesner, L.A. and T. Paul (1994), "Overview Of Federal Law And USEPA Regulations For Urban Runoff," Water Science and Technology, Vol. 37, No. 1, Elsevier Science Ltd.

Sakakibara, T. (1996), "Sediments Flushing Experiment in a Trunk Sewer," Water Science and Technology, Vol. 33, No. 9, International Association for Water Quality (IAWQ), United Kingdom.

Schmidt, J., P. Seto, and D. Averill (1997), "Pilot-Scale Study of Satellite Treatment Options for Control of Combined Sewer Overflows," Water Quality Research Journal Canada, Vol. 32, No. 1, Canadian Association for Water Quality (CAWQ), Canada.

Schnell, G. (1991), "Investigation of Nitrification Under Storm Conditions in the Main Treatment Plant, Vienna," Water Science and Technology, Vol. 24, No. 7, International Association for Water Quality (IAWQ), United Kingdom.

Seidl, M., P. Servais, M. Martaud, C. Gandouim, and J.M. Mouchel (1998), "Organic Carbon Biodegradability and Heterotrophic Bacteria Along A Combined Sewer Catchment During Rain Events," Water Science and Technology, Vol. 37, No. 1, Elsevier Science Ltd.

Shao, J., A. Liu, F. Wada, J. Crosse, and D. Jenkins (1996), "Advanced Primary Treatment: An Alternative to Biological Secondary Treatment. The City of Los Angeles Hyperion Treatment Plant Experience," Water Science and Technology, Vol. 34, No. 3-4, International Association for Water Quality (IAWQ), United Kingdom.

Sorber, C. (1983), "Chapter 3 – Removal of Viruses from Wastewater and Effluent by Treatment Processes," Viral Pollution of the Environment, Published by CRC Press Inc, Boca Raton, Fla.

Stallard, M., W. Smith, R. Crites, G. Tchobanoglous (1981), "Design of Storage/Sedimentation Facilities to Control Urban Runoff and Combined Sewer Overflows," in Restoration of Lakes and Inland Waters, EPA Document EPA-440/5-81-010, U.S. Environmental Protection Agency, Office of Water, Washington, DC.

Stephenson, J., B. Gall, M., Mroczek, M., Newbigging, and J. Parker (2002), "Assessment of Technologies for Screening, Floatable Control, and Screenings Handling," Project No. 00-CTS-4, Water Environment Research Foundation, Alexandria, Va.

Sullivan, R. et al. (1982), "Swirl and Helical Bend Pollution Control Devices," EPA Document EPA 600/8-82-013, U.S. Environmental Protection Agency, Office of Research and Development, Cincinnati, Ohio.

Tchobanoglous, G. (1999), "The Impact of Particles in Wastewater on UV Disinfection," Proceedings of the 5th EPA National Wastewater Treatment Technology Transfer Workshop, EPA Document EPA 907/R-98/001, U.S. Environmental Protection Agency, Office of Research and Development, Washington, DC.

Urbinato, David (1994), "Portland, Maine: Case of a Combined Sewer System Subhead: When it rains, the system overflows," EPA Journal, Clean Water Agenda, Summer. Internet: http://www.epa.gov/docs/epajrnal/summer94/.

Vaes, G. and J. Berlamont (1999), "The impact of rainwater reuse on CSO emissions," Water Science and Technology, Vol. 39, No. 5, International Association for Water Quality (IAWQ), United Kingdom.

Vanderborght, J.P. and P. Wollast (1990) "Continuous Monitoring of Wastewater Compisition in Sewers and Stormwater Overflows," Water Science and Technology, Vol. 22, No. 10-11, International Association for Water Quality (IAWQ), United Kingdom.

Vanrolleghem, P., U. Jeppsson, J. Carstensen, B. Carlsson, and Gustaf Olsson (1996), "Integration of Wastewater Treatment Plant Design and Operation – A Systematic Approach Using Cost Functions," Water Science and Technology, Vol. 34, No. 3-4, International Association for Water Quality (IAWQ), United Kingdom.

WEF (1991), "Design of Municipal Wastewater Treatment Plants: Vol.s I and II," WEF Manual of Practice No. 8, Water Environment Federation, Alexandria, Va.

WEF (1997), "Automated Process Control Strategies," Water Environment Federation, Alexandria, Va.

West, R., R. Sawey, and D. Gerrity (2000), "Pilots Show Alternative Processes Manage Storms Efficiently at Less Cost," Watershed & Wet Weather Technical Bulletin, Vol. 5, No. 4, October 2000, Water Environment Federation, Alexandria, Va.

Wong, T.H.F. and W.F. Geiger (1997), "Adaptation of Wastewater Surface Flow Wetland Formulae For Application In Constructed Stormwater Wetlands," Ecological Engineering, Vol. 9, No. 3-4.

2002 WERF Subscribers

The following agencies and corporations are WERF's charter stormwater research supporters:

Stormwater Agency

California
Monterey, City of
Santa Rosa, City of

Colorado
Boulder, City of

Georgia
Griffin, City of

Kansas
Overland Park, City of

Kentucky
Louisville & Jefferson County Metropolitan Sewer District

North Carolina
Cary, Town of
Charlotte, City of/Mecklenburg County

Pennsylvania
Philadelphia, City of

Corporate

BaySaver Inc.
Vortechnics Inc.
WWETCO, LLC

State

Fresno Metropolitan Flood Control District, Calif.
Urban Drainage & Flood Control District, Colo.

Wastewater Utility

Alabama
Montgomery Water Works & Sanitary Sewer Board

Alaska
Anchorage Water & Wastewater Utility

Arizona
Gila Resources
Glendale, City of, Utilities Department
Mesa, City of
Phoenix Water Services Department
Pima County Wastewater Management

Arkansas
Little Rock Wastewater Utility

California
Central Contra Costa Sanitary District
Contra Costa Water District
Crestline Sanitation District
Delta Diablo Sanitation District
Dublin San Ramon Services District
East Bay Municipal Utility District
El Dorado Irrigation District
Escondido, City of
Fairfield-Suisun Sewer District
Irvine Ranch Water District
Las Virgenes Municipal Water District
Lodi, City of
Los Angeles, City of
Los Angeles County, Sanitation Districts of
Orange County Sanitation District
Palo Alto, City of
Riverside, City of
Sacramento Regional County Sanitation District
San Diego Metropolitan Wastewater Department, City of
San Francisco, City & County of
San Jose, City of
Santa Barbara, City of
Santa Rosa, City of
South Bayside System Authority
South Orange County Wastewater Authority
Union Sanitary District

Colorado
Boulder, City of
Colorado Springs, City of
Littleton/Englewood Water Pollution Control Plant
Metro Wastewater Reclamation District, Denver

Connecticut
New Haven, City of, WPCA

District of Columbia
District of Columbia Water & Sewer Authority

Florida
Broward, County of
Fort Lauderdale, City of
Gainesville Regional Utilities
JEA
Kissimmee, City of, Department of Water Resources
Miami-Dade Water & Sewer Authority
Orange County Utilities Department
Orlando, City of
Reedy Creek Improvement District
Seminole County Environmental Services
St. Petersburg, City of
Stuart Public Utilities
Tallahassee, City of
Tampa, City of
West Palm Beach, City of

Georgia
Atlanta Wastewater Services
Augusta, City of
Clayton County Water Authority
Cobb County Water System
Columbus Water Works
Fulton County
Gwinnett County Department of Public Utilities
Macon Water Authority
Savannah, City of

Hawaii
Honolulu, City and County of

Illinois
American Bottoms Wastewater Treatment Plant
Dupage, County of
Greater Peoria Sanitary District
Metropolitan Water Reclamation District of Greater Chicago
Wheaton Sanitary District

Indiana
Fort Wayne, City of
Indianapolis, City of

Iowa
Ames, City of
Des Moines Metro Wastewater Reclamation Authority

Kansas
Johnson County Unified Wastewater Districts
Unified Government of Wyandotte County/Kansas City, City of

Kentucky
Louisville & Jefferson County Metropolitan Sewer District

Louisiana
Sewerage & Water Board of New Orleans

Maine
Bangor, City of

Maryland
Anne Arundel County Bureau of Utility Operations
Howard County Department of Public Works
Washington Suburban Sanitary Commission

Massachusetts
Boston Water & Sewer Commission
Massachusetts Water Resources Authority
Upper Blackstone Water Pollution Abatement District

Michigan
Ann Arbor, City of
Detroit, City of

Grand Rapids, City of
Holland Board of Public Works
Lansing, City of
Owosso Mid-Shiawassee County WWTP
Saginaw, City of
Wayne County Department of Environment
Wyoming, City of

Missouri
Independence, City of
Kansas City Missouri Water Services Department
Little Blue Valley Sewer District
Metropolitan St. Louis Sewer District

Nebraska
Lincoln Wastewater System

Nevada
Henderson, City of

New Jersey
Passaic Valley Sewerage Commissioners

New York
New York City Department of Environmental Protection

North Carolina
Cary, Town of
Charlotte/Mecklenburg Utility Department
Durham, City of
Metropolitan Sewerage District of Buncombe County
Orange Water & Sewer Authority

Ohio
Akron, City of
Butler County Department of Environmental Services
Columbus, City of
Metropolitan Sewer District of Greater Cincinnati
Northeast Ohio Regional Sewer District

Oklahoma
Tulsa, City of

Oregon
Clean Water Services
Eugene/Springfield Water Pollution Control
Water Environment Services

Pennsylvania
Philadelphia, City of
University Area Joint Authority, State College

South Carolina
Charleston Commissioners of Public Works
Mount Pleasant Waterworks & Sewer Commission
Spartanburg Sanitary Sewer District

Tennessee
Cleveland, City of
Knoxville Utilities Board
Murfreesboro Water & Sewer Department
Nashville Metro Water Services

Texas
Amarillo, City of
Austin, City of
Dallas Water Utilities
Denton, City of
El Paso Water Utilities
Fort Worth, City of
Gulf Coast Waste Disposal Authority
Houston, City of
San Antonio Water System
Trinity River Authority

Utah
Salt Lake City Corporation

Virginia
Alexandria Sanitation Authority
Arlington, County of
Fairfax County Virginia
Hampton Roads Sanitation District
Henrico, County of
Hopewell Regional Wastewater Treatment Facility
Loudoun County Sanitation Authority
Prince William County Service Authority
Richmond, City of

Washington
Edmonds, City of
Everett, City of
King County Department of Natural Resources
Seattle Public Utilities

Wisconsin
Green Bay Metro Sewerage District
Madison Metropolitan Sewerage District
Milwaukee Metropolitan Sewerage District
Racine, City of
Sheboygan Regional Wastewater Treatment
Wausau Water Works

Australia
Sydney Water Corporation

Canada
Toronto, City of, Ontario

Mexico
Servicios de Agua y Drenaje de Monterrey, I.P.D.

United Kingdom
Yorkshire Water Services Limited

Corporate

The ADVENT Group Inc.
Alan Plummer & Associates
Alpine Technology Inc.
Aquateam–Norwegian Water Technology Centre A/S
Black & Veatch
Boyle Engineering Corporation
BPR CSO
Brown & Caldwell
Burns & McDonnell
The Cadmus Group
Camp Dresser & McKee Inc.
Carollo Engineers Inc.
Carpenter Environmental Associates Inc.
CDS Technologies Inc.
Chemtrac Systems Inc.
CH2M HILL
Clancy Environmental Consultants Inc.
Damon S. Williams Associates, LLC
Earth Tech Inc.
Ecolab Water Care Services
EMA Inc.
The Eshelman Company Inc.
Finkbeiner, Pettis, & Strout Inc.
Frontier Geosciences Inc.
ftn Associates Inc.
Gannett Fleming Inc.
GE Betz
Golder Associates Inc.
Greeley and Hansen LLC
The HACH Company
Hazen & Sawyer, P.C.
HDR Engineering Inc.
HNTB Corporation
HydroQual Inc.
Insituform Technologies Inc.
Institute for Environmental Technology & Industry, Korea
Jacobson Helgoth Consultants Inc.
Jason Consultants Inc.
Jordan, Jones, & Goulding Inc.
KCI Technologies Inc.
Kelly & Weaver, P.C.
Kennedy/Jenks Consultants
Komline Sanderson Engineering Corporation
Lawler, Matusky & Skelly Engineers, LLP
Limno-Tech Inc.
Lombardo Associates Inc.
Malcolm Pirnie Inc.
McKim & Creed
MWH
Odor & Corrosion Technology Consultants Inc. (OCTC)
ONDEO Degremont Inc.
PA Government Services Inc.
Parametrix Inc.
Parsons
Post, Buckley, Schuh & Jernigan
The RETEC Group
R.M. Towill Corporation
Royce Technologies
Sear-Brown
SYNAGRO
Tetra Tech Inc.
Trojan Technologies Inc.
URS Corporation
USFilter NATC
Wade-Trim Inc.
Weston Solutions Inc.
Woodard & Curran
Woodruff & Howe Environmental Engineering Inc. (WHEE)
WRc/D&B, LLC

Industry

American Electric Power
BP-Amoco Corporation
Chevron Research & Technology Company
The Coca-Cola Company
Dow Chemical Company
DuPont Company
Eastman Chemical Company
Eastman Kodak Company
Merck & Company Inc.
ONDEO Services
Procter & Gamble Company
Reliant Energy
Severn Trent Services Inc.
Shell Global Solutions
Thames Water Plc
United Water Services LLC

WERF Product Order Form

As a benefit of joining the Water Environment Research Foundation, subscribers are entitled to receive one complimentary copy of all final reports and other products. Additional copies are available at cost (usually $10). To order your complimentary copy of a report, please write "free" in the unit price column. WERF keeps track of all orders. If the charge differs from what is shown here, we will call to confirm the total before processing.

Name　　Title

Organization

Address

City　　State　　Zip Code　　Country

Phone　　Fax　　Email

Stock #	Product	Quantity	Unit Price	Total
			Postage & Handling	
			VA Residents Add 4.5% Sales Tax	
			Canadian Residents Add 7% GST	
			TOTAL	

Method of Payment:

(All orders must be prepaid.)

❑ Check or Money Order Enclosed

❑ Visa　❑ Mastercard　❑ American Express

Account No.　　Exp. Date

Signature

Shipping & Handling:

Amount of Order	United States	Canada & Mexico	All Others
Up to but not more than:	Add:	Add:	Add:
$20.00	$5.00*	$8.00	50% of amount
30.00	5.50	8.00	40% of amount
40.00	6.00	8.00	
50.00	6.50	14.00	
60.00	7.00	14.00	
80.00	8.00	14.00	
100.00	10.00	21.00	
150.00	12.50	28.00	
200.00	15.00	35.00	
More than $200.00	Add 20% of order	Add 20% of order	

*minimum amount for all orders

Note: Please make checks payable to the Water Environment Research Foundation.

To Order (Subscribers Only):

Log on to www.werf.org and click on the "Product Catalog."

Phone: (703) 684-2470
Fax: (703) 299-0742.

WERF
Attn: Subscriber Services
601 Wythe Street
Alexandria, VA 22314-1994

To Order (Non-Subscribers):

Non-subscribers may be able to order WERF publications either through WEF (www.wef.org) or IWAP (www.iwap.co.uk). Visit WERF's website at www.werf.org for details.